NO
FILTER

The small Instagram team at Facebook headquarters in 2015
by John Barnett

THE
INSIDE
STORY
OF HOW
INSTAGRAM
TRANSFORMED
BUSINESS,
CELEBRITY
AND OUR
CULTURE

NO
FILTER

SARAH FRIER

1 3 5 7 9 10 8 6 4 2

Random House Business
20 Vauxhall Bridge Road
London SW1V 2SA

Random House Business is part of the Penguin Random House
group of companies whose addresses can be found
at global.penguinrandomhouse.com.

First published in the United Kingdom by Random House Business in 2020
First published in the United States by Simon & Schuster in 2020

www.penguin.co.uk

A CIP catalogue record for this book is available from the British Library

ISBN 9781847942524 (Hardcover)
ISBN 9781847942531 (Trade paperback)

Interior design by Lewelin Polanco

Printed and bound in Great Britain by Clays Ltd, Elcograf S.p.A.

Penguin Random House is committed to a sustainable future for
our business, our readers and our planet. This book is made
from Forest Stewardship Council® certified paper.

To Matt

CONTENTS

Author's Note   xi

Introduction: The Ultimate Influencer   xv

1 | PROJECT CODENAME   1

2 | THE CHAOS OF SUCCESS   30

3 | THE SURPRISE   50

4 | THE SUMMER IN LIMBO   69

5 | MOVE FAST AND BREAK THINGS   87

6 | DOMINATION   106

7 | THE NEW CELEBRITY   126

8 | THE PURSUIT OF THE INSTA-WORTHY   159

9 | THE SNAPCHAT PROBLEM   179

10 | CANNIBALIZATION   207

11 | THE OTHER FAKE NEWS   229

12 | THE CEO   252

Epilogue: The Price of the Acquisition   275

Acknowledgments   281

Notes   287

Index   303

AUTHOR'S NOTE

This book is an effort to bring you the definitive inside story of Instagram. It would not have been possible without the hundreds of people—current and former employees, executives and others who built their careers around the app, as well as competitors—who volunteered their time and shared memories they've never shared with a journalist before. Instagram's founders spoke with me both together and separately over several years. Facebook Inc. offered more than two dozen sit-down interviews with current staff and executives, including the current head of Instagram, even after the founders departed their company.

Despite tensions between the founders and their acquirer, and despite the flurry of critical stories I was writing about Facebook as a journalist for Bloomberg News, everyone agreed it was important to have this book be as accurate as possible. When potential sources forwarded my outreach to either the founders or the company, asking if it was okay to talk to me, they were generally told yes, even though both the founders and the company knew they would have no control over the ultimate content of this book. I commend them for that decision.

Still, most of the sources for the book talked without explicit permission, or the company's knowledge. When they talked with me, they risked violating the strict nondisclosure agreements that employees sign when they join. In fact, every non-journalist who visits Facebook's headquarters has to sign a nondisclosure contract when they enter through security, before they are allowed to meet with an employee. For that reason, most of my sources provided their interviews, documents, and other materials only anonymously.

That context is important for understanding why I wrote the book the way I did: in a narrative style, presenting the story through an omniscient perspective that incorporates all these different memories. I do not directly say who told me what information, in order to protect my sources. In areas where I build off existing news reports, I have cited the prior reporting in my endnotes. I made the choice to quote from on-record interviews only when I'm bringing in an outsider, like a celebrity or an influencer whose perspective enriches our understanding of the app's impact on the world.

Since starting this project, I asked for and hoped to receive an interview with Mark Zuckerberg for this book. I argued that the Facebook CEO, whom I'd interviewed several times in previous years and whom I watched testify in front of the U.S. Congress for ten hours in 2018, has become something of a villain in our public imagination. A book like this, I told a public relations representative, is an opportunity to look at all those important moments we've written about in Facebook's history, and dig into everything we didn't fully understand when it happened.

There were many hard questions to ask, but I could start with an easy one. Why did Zuckerberg want to buy Instagram? Not the answer in his blog post, but the personal story. What were the steps and triggers that caused him to decide, on a Thursday in April 2012, that he needed to pick up his phone and start a process to acquire the company *as soon as possible*? And not just buy it, but commit to keeping it independent?

One month before this manuscript was due, I received an email from Facebook public relations, with an answer to that question, attributable to Zuckerberg:

"It's simple. It was a great service and we wanted to help it grow."

That's the whole quote on the matter. To bring you the full story, then, I have relied on others to remember what Zuckerberg said in key moments, or what he was thinking, based on what he told peers. I have checked those recollections with Facebook, though they generally chose not to comment on such anecdotes.

In general, readers cannot assume that people speaking in this book have provided that exact dialogue to me. In most cases, a person in a given conversation told me what they said from their own memory. But sometimes others have remembered the details better. I write dialogue exactly as it was relayed to me in interviews, in an attempt to show Instagram's journey as its participants remember it. But my sources, even those who remember their own thoughts and words, may remember them in a simplified form, or incorrectly, or in a way that contradicts other sources, because the Instagram story takes place over ten years. This book is my best attempt to provide the truth of the Instagram story, with no filter except my own.

INTRODUCTION

THE ULTIMATE INFLUENCER

In São Paulo, Brazil, there is an open-air gallery of street art called Beco do Batman, or Batman's Alley. Its nickname long preceded the creation of one of its more memorable murals, which depicts, in 17 feet of chipped paint, the Brazilian soccer legend Pelé in an embrace with the Dark Knight. We only know it's Pelé because of the no. 10 jersey with his name; otherwise he is facing away, cheek pressed up against Batman's mask, perhaps giving a kiss or telling a secret, while Batman's hand grazes Pelé's lower back.

On a Saturday in March, a young woman stands in front of the mural, about the height of the number on Pelé's jersey. She looks intentionally casual, in sunglasses, red sneakers, and a loose white top. Her friend snaps images of her smiling, and then a few more of her with a contemplative, distant stare. They move on to the next mural, and the next, waiting patiently for a turn with the more popular backdrops. Dozens of others are doing the same, including three soon-to-be moms in crop tops, who have friends along to document the size of their baby bulges in front of a surreal purple orchid. Nearby a blonde little girl in sequined blue-and-red shorts, red lipstick, and a shirt that says "Daddy's Little Monster" is holding a baseball bat and posing in front of an ominous bird mural; her mother instructs her to hold the bat higher, more fiercely, to look more like Harley Quinn from the *Suicide Squad* comics. She obliges.

Along the curve of the alley, vendors take advantage of the crowd, selling beer and jewelry. A man strums a guitar and sings in Portuguese, hoping to build a fan base for his music. On his instrument, he's taped a large piece of paper with the name of his social media account, alongside the logo of the only app that matters here: Instagram.

With the rise of Instagram, Beco do Batman has become one of São Paulo's top tourist destinations. Via the vacation rental site Airbnb, various vendors charge about $40 per person to provide two hours of "personal paparazzi" in the alley, taking high-quality pictures of people to post on Instagram; the service is a type that's become one of Airbnb's most popular for its travelers in cities around the world.

For amateur photographers, the only cost is the stress of perfection. One woman corrals two small children sparring over a bottle of Coca-Cola so that her sister can stand in line to pose in front of green-and-blue peacock feathers. The teenager who just had her turn with the peacock gets angry with her companion for wasting it by snapping an unflattering angle. But nobody photographs the photographers; on Instagram, the polished images become the reality, driving more and more visitors to this place.

I came to the alley on a recommendation from a man named Gabriel, whom I happened to sit next to at a sushi bar my first night in Brazil. My Portuguese language skills were so poor that he stepped in to translate for the restaurant workers. I explained that I was on a journey to understand more about Instagram and its impact on culture around the world. As we talked, and as the chef delivered bites of sashimi and nigiri, Gabriel photographed each dish to post on his Instagram story, while lamenting that his friends were so obsessed with sharing their lives, he wasn't sure if they were actually *living* them.

• • • • • • • • • • • •

Each month, more than 1 billion of us use Instagram. We take photos and videos of our food, our faces, our favorite scenery, our families, and our interests and share them, hoping that they reflect something about who we are or who we aspire to be. We interact with these posts and each other, aiming to forge deeper relationships, stronger networks, or personal brands. It's just the way modern life works. Rarely do we have the chance to reflect on how we got here and what it means.

But we should. Instagram was one of the first apps to fully exploit our relationship with our phones, compelling us to experience life through a camera for the reward of digital validation. The story of Instagram is an overwhelming lesson in how the decisions inside a social media company—about what users listen to, which products to build, and how to measure success—can dramatically impact the way we live, and who is rewarded in our economy.

I aim to take you behind the scenes with cofounders Kevin Systrom and Mike Krieger as they navigated what to do with their product's power over our attention. Every decision they made had a dramatic ripple effect. By selling their company to Facebook, for example, they ensured Instagram's longevity while helping the social media giant become even more powerful and formidable versus its competitors. After the sale, the Instagram founders became disillusioned with Facebook's utilitarian grow-at-all-costs culture and resisted it, focusing instead on building a thoughtfully crafted product, where what's popular is shaped by Instagram's own storytelling about its biggest users. The plan worked so well that Instagram's success ended up threatening Facebook and its CEO, Mark Zuckerberg.

The way the story ended for the Instagram founders, with their tense departure from the company in 2018, is not the way it will end for the rest of us. Instagram is now so entangled with our daily lives that the business story cannot be detached from its impact on us. Instagram has become a tool with which to measure cultural relevance, whether it's in a school, in an interest-based community, or in the world. A substantial portion of our global population is striving for digital recognition and validation, and many of them are getting it through likes, comments, followers, and brand deals. Inside and outside Facebook, the story of Instagram is ultimately about the intersection of capitalism and ego—about how far people will go to protect what they built and to appear successful.

The app has become a celebrity-making machine the likes of which the world has never seen. More than 200 million of Instagram's users have more than 50,000 followers, the level at which they can make a living wage by posting on behalf of brands, according to the influencer analysis company Dovetale. Less than a hundredth of a percent of Instagram's users have more than a million followers. At Instagram's massive scale, that 0.00603 percent equates to more than 6 million Insta-celebrities, a majority of them rising to fame through the app itself. For a sense of scale, consider that millions of people and brands have more Instagram followers than the *New York Times* has subscribers. Marketing through these people, who are basically running personal media companies through tastemaking, storytelling, and entertaining, is now a multibillion-dollar industry.

All of this activity has trickled down into our society, affecting us whether we use Instagram or not. Businesses that want our attention—from hotels and restaurants to large consumer brands—change the way they design their spaces and how they market their products, adjusting their strategies to cater to the new visual way we communicate, to be worthy of photographing for Instagram. By looking at the way commercial spaces, products, and even homes are designed, we can *see* Instagram's impact, in a way that we can't as easily see the impact of Facebook or Twitter.

The San Francisco workspace I'm writing this book in, for example, has its library arranged not according to title or author, but cover color: the decision makes sense when prioritizing Instagram aesthetics over book discovery. A burger chain founded in Manhattan called Black Tap created indulgent milkshakes with entire slices of cake on top, and for months people lined up around the block to purchase them. Even though diners rarely finished their mega-desserts, they felt compelled to photograph them. In Japan, there is a word for this Insta-grammable design movement: *Insta-bae* (インスタ映え), pronounced "Insta-bye-eh." The more Insta-bae something is, whether it's an outfit or a sandwich, the more socially and commercially successful it has the potential to be.

I talked to a university student in London who explained that a higher Instagram follower count means you're more likely to be selected for a leadership role on campus. I talked to a woman in Los Angeles who is too young to drink legally but gets called up by club promoters to attend exclusive events because of her sizable Instagram following. I talked to a parent in Indonesia whose daughter goes to school in Japan, then brings back Japanese consumer goods in suitcases every summer to sell locally by posting photos of the products on Instagram. I talked to a Brazilian couple who built an entire baking business out of their apartment kitchen, drawing tens of thousands of followers, because their doughnuts come in the shape of letters that say "I love you!"

Instagram has fueled careers and even celebrity empires. Kris Jenner, the manager of the Kardashian-Jenner reality television family, says Instagram has transformed the job beyond the show *Keeping Up with the Kardashians* and into a 24/7 content and brand promotion cycle. She wakes up between 4:30 and 5 a.m. in her palatial Hidden Hills, California, home and checks Instagram before she does anything else. "I literally can go on Instagram and check my family, my grandchildren, my business," she explained. "I just immediately check on my kids. What's everybody doing? Are they awake? Are they posting pictures for the business on schedule? Are they having fun?"

The Instagram schedule is posted in Kris's office, but also on a printout she receives every night and every morning. She and her children represent dozens of brands between them, including Adidas, Calvin Klein, and Stuart Weitzman, but are also launching their own makeup and beauty lines. The family's five sisters—Kim Kardashian West, Kylie Jenner, Kendall Jenner, Khloé Kardashian, and Kourtney Kardashian—have a combined reach of more than half a billion followers.

The day we speak, Kris is en route to an Instagrammable pink-themed party to launch a skin-care line for her daughter Kylie. She recalls the first time Kylie asked if it was okay to start a lipstick business, just through her Instagram feed, without any physical product in stores. "I said to her, 'You're going to start with three colors in your lip kit and they've got to be colors you really love,'" Kris remembers. "So either it's going to be amazing and fly off the shelf, or it's going to flop and you're going to be wearing these three colors for the rest of your life."

They were together in Kris's office in 2015 when Kylie posted the link to the website. Within seconds, the entire product was sold out. "I thought something went wrong," Kris recalls. "Did this break? Did the website crash? What happened?"

It was not a fluke. It was just an indication that whatever her daughter would tell people to do, they would do. Over the next few months, whenever Kylie would announce on her Instagram that new products were coming, there would be more than 100,000 people waiting on her website for them to drop. Four years later, when Kylie was 21, *Forbes* put her on its cover and declared her the youngest "self-made" billionaire ever. Now every beauty guru on Instagram seems to have his or her own product line.

• • • • • • • • • • • •

There is something powerful about that number—1 billion—in our society. It's a marker signifying, especially in business, that you've achieved some unique untouchable status, graduating into an echelon that inspires awe and merits newsworthiness. In 2018, when *Forbes* published a story putting Jenner's net worth just shy of that threshold at $900 million, Josh Ostrovsky, the owner of the popular and controversial Instagram humor account @thefatjewish, told his followers to donate to a crowdfunding campaign to raise $100 million for Kylie. "I don't want to live in a world where Kylie Jenner doesn't have a billion dollars," he wrote in his caption, spurring a tongue-in-cheek viral news cycle.

After being acquired by Facebook, in a deal that shocked the industry, Instagram became the first-ever mobile app to achieve a $1 billion valuation. Instagram's success was unlikely, as is that of all startups. When it launched in 2010, the app didn't start out as a popularity contest, or as an avenue for personal branding. It caught on because it was a place to see into someone else's life and how they experienced it through their phone's camera.

According to Chris Messina, the technologist who was user no. 19 and invented the hashtag, the introduction to other people's visual perspectives on Instagram was a stunning novelty—perhaps equivalent to the psychological phenomenon astronauts experience when looking at the Earth from outer space for the first time. On Instagram, you could dive into the life of a reindeer herder in Norway or a basket weaver in South Africa. And you could share and reflect on your own life in a way that felt profound too.

"It gives you this glimpse of humanity and changes your whole perspective on everything and the importance of it," Messina explained. "Instagram is this mirror on ourselves, and it allows each of us to contribute our own experience to the understanding of this world."

As Instagram grew, its founders tried to preserve this sense of discovery. They became aesthetic tastemakers for a generation, responsible for imbuing us with a reverence for visually arresting experiences that we can share with our friends and strangers for the reward of likes and followers. They invested heavily in an editorial strategy to show how they intended Instagram to be used: as a venue for different perspectives and creativity. They eschewed some of Facebook's spammy tactics, like sending excessive notifications and emails. They resisted adding tools that would have helped fuel the influencer economy. You can't add a hyperlink in a post, for example, or share someone's post the way you can on Facebook.

And until recently, they never changed the measurements to make it possible for us to compare ourselves to each other and try to ascend to higher levels of relevance. In the app, Instagram gave its users three simple measures for how they were performing: a "follower" count, a "following" count, and "likes" on their photographs. These feedback scores were enough to make the experience thrilling, even addicting. With every like and follow, an Instagram user would get a little rush of satisfaction, sending dopamine to the brain's reward centers. Over time, people figured out how to be good at Instagram, unlocking status in society and even commercial potential.

And because of filters that initially improved our subpar mobile photography, Instagram started out as a place for enhanced images of people's lives. Users began to accept, by default, that everything they were seeing had been edited to look better. Reality didn't matter as much as aspiration and creativity. The Instagram community even devised a hashtag, #nofilter, to let people know when they were posting something raw and true.

The account with the largest following on Instagram, at 322 million, is @instagram, the one controlled by the company. It's fitting, because Instagram holds the utmost influence over the world it has shaped. In 2018, Instagram reached 1 billion monthly users—their second "1 billion" milestone. Soon after, the founders left their jobs. As Systrom and Krieger discovered, even if you reach the highest echelons of business success, you don't always get what you want.

NO
FILTER

PROJECT CODENAME

"I like to say I'm dangerous enough to know how to code and sociable enough to sell our company. And I think that's a deadly combination in entrepreneurship."

—KEVIN SYSTROM, INSTAGRAM COFOUNDER

K evin Systrom had no intention of dropping out of school, but he wanted to meet with Mark Zuckerberg anyway.

Systrom, who is six-foot-five with dark brown hair, squinty eyes, and a rectangular face, had met this local startup founder through Stanford University friends earlier in 2005, while sipping beers in red plastic cups at a party in San Francisco. Zuckerberg was becoming a tech industry wunderkind for his work on TheFacebook.com, a social network he'd started with friends the year prior at Harvard University and then expanded to college campuses around the country. Students used the website to write short updates about what they were doing, then post those statuses on their Facebook "walls." It was just a simple site, with a white background and blue trim, not like the social network Myspace, with its loud designs and customizable fonts. It was also growing so fast that Zuckerberg decided there was no reason to go back to school.

At Zao Noodle Bar on University Avenue, about a mile from Stanford's campus, Zuckerberg tried to convince Systrom to make the same decision. They were both just past legal drinking age but Zuckerberg—about ten inches shorter than Systrom, with light curly hair and pale pink skin, always wearing slip-on Adidas sandals, loose-fitting jeans, and a zip-up hoodie—looked much younger. He wanted to add photos to the Facebook experience, beyond the singular profile picture, and wanted Systrom to build the tool.

Systrom was pleased to be recruited by Zuckerberg, who he thought was hyperintelligent. He did not consider himself a stellar programmer. At Stanford, he felt like a regular person among prodigies from around the world, and barely scraped up a B in his first and only computer science course. Still, he fit the general category of what Zuckerberg needed. He did like photography, and one of his side projects was a website called Photobox, which allowed people to upload large image files and then share or print them, especially after parties at his fraternity, Sigma Nu.

Photobox was enough to interest Zuckerberg, who wasn't exactly picky at this point. Recruiting is always the hardest part of building a startup, and TheFacebook.com was growing so fast that he needed bodies in the room. Earlier that year, Zuckerberg was spotted in front of Stanford's computer science building holding up a poster about his company, hoping to nab coders the way campus clubs recruited members. He had nailed the pitch, explaining to Systrom that he was offering a once-in-a-lifetime chance to be on the ground floor of something that would be truly huge. Facebook was going to open up to high school students next, and eventually to the whole world. The company was going to raise more money from venture capitalists and might one day be bigger than Yahoo! or Intel or Hewlett-Packard.

And then, when the restaurant ran Zuckerberg's credit card, it didn't go through. He blamed it on the company's president, Sean Parker.

A few days later, Systrom went on a walk in the foothills near campus with the assigned mentor from his entrepreneurship program—a 1978 Stanford MBA in venture investing named Fern Mandelbaum. She worried Systrom would waste his potential if he gave everything up for somebody else's vision. "Don't do this Facebook thing," she said. "It's a fad. It won't go anywhere."

Systrom thought she was right. Either way, he hadn't come to Silicon Valley to get rich quick from a startup. He intended to get a world-class education and graduate from Stanford. He thanked Zuckerberg for his time, then planned for a different kind of adventure: studying abroad in Florence, Italy. They would keep in touch.

• • • • • • • • • • • • •

Florence spoke to Systrom in a way TheFacebook.com didn't. Systrom wasn't sure he was supposed to work in technology. When he'd first applied to Stanford, he'd thought he would major in structural engineering and art history. He imagined traveling the world, restoring old cathedrals or paintings. He loved the science behind art, and how a simple innovation—like architect Filippo Brunelleschi's rediscovery of linear perspective during the Renaissance—could completely change the way people communicated. The paintings for most of Western history were flat and cartoonish, and then, starting in the 1400s, perspective gave them depth, making them photorealistic and emotive.

Systrom liked thinking about the way things were made, decoding the systems and details that mattered for producing something of quality. In Florence, he developed mini-obsessions with Italian crafts, learning the steps in wine making, the process for shaping and stitching leather for shoes, and the techniques for spinning up a legitimately good cappuccino.

Even during his charmed childhood, Systrom would explore his hobbies with this level of academic fervor, in pursuit of perfection. Born in December 1983, he was raised, along with his sister, Kate, in a two-story house with a long driveway on a tree-lined street in suburban Holliston, Massachusetts, about an hour west of Boston. His energetic mother, Diane, was vice president of marketing at nearby Monster.com, and later at Zipcar, and introduced her children to the internet back when the connection took over the phone line. His father, Doug, was a human resources executive at the conglomerate that owned Marshalls and HomeGoods discount stores. Systrom was an earnest, curious kid who loved going to the library and playing the futuristic, demon-riddled first-person shooter game *Doom II* on the computer. His introduction to computer programming was creating his own levels in the game.

He would jump from intense passion to intense passion, in phases that everyone around him had to hear about—sometimes quite literally. During his deejaying phase at a Middlesex boarding school, he bought two turntables and stuck an antenna out his dorm room window to broadcast his own radio station, playing electronic music, which was niche at the time. As a teen, he would sneak into 21-and-over clubs to observe his idols in action, but he was too rule-abiding to drink there.

People either instantly loved Systrom or wrote him off as pretentious and superior, someone who put on airs. He was good at listening to others, but was also quite willing to teach people about the right way to do things, eliciting, because his obsessions were so varied, either fascination or eye rolls. He was the kind of person who would say that he wasn't good at something he was actually good at, or that he wasn't cool enough to do something he was actually cool enough to do, toeing the line between being relatable and humblebragging. For instance, to fit in in Silicon Valley, he would often note his high school nerd credentials—his video gaming and coding side projects—but rarely mentioned that he'd also been the captain of the lacrosse team or that he was in charge of hyping parties for his college fraternity. His frat brothers considered him innovative for using viral video as a means of summoning attendees in the thousands. Systrom's first such production, in 2004, was called Moonsplash and featured fraternity brothers dancing in off-color costumes to Snoop Dogg's "Drop It Like It's Hot." Systrom would always deejay at the events.

Photography was one of his longest-running personal interests. He wrote, for a class in high school, that he liked using the medium "to show my outlook on the world to everyone" and "inspire others to look at the world in a new way." Ahead of his trip to Florence, the epicenter of the Renaissance he'd learned so much about, he saved up to purchase, after intensive research, one of the highest-quality cameras he could afford, with the sharpest lens. He intended to use it in his photography class.

His teacher in Florence, a man named Charlie, was unimpressed. "You're not here to do perfection," he said. "Give me that."

Systrom thought the professor was going to change the settings on the camera. Instead, he took the prized purchase away into his back room and returned with a smaller device, called a Holga, that only took blurry, square black-and-white photographs. It was plastic, like a toy. Charlie told Systrom he wasn't allowed to use his fancy camera for the next three months, because a higher-quality tool wouldn't necessarily create better art. "You have to learn to love imperfection," he instructed.

Systrom spent the winter of his junior year, in 2005, snapping photos here and there in cafes, trying to appreciate a blurry, out-of-focus beauty. The idea—of a square photo transformed into art through editing—stuck in the back of Systrom's mind. More important was the lesson that just because something is more technically complex doesn't mean it's better.

Meanwhile, he was making plans for the summer ahead. Systrom needed a startup internship as part of the Stanford Mayfield Fellows Program he'd been barely accepted into. Like all Stanford students, he'd had a front-row seat to the resurrection of the internet industry. The first generation of the web was about moving information and businesses online, fueling a speculative dot-com gold rush boom in the late 1990s that busted in 2001. This new generation, which investors separated from the failures with the jargony term "Web 2.0," was about making websites more interactive and interesting, relying on information their users created, like restaurant reviews and blogs.

Most of the hot new tech was in suburban Palo Alto, where companies with names like Zazzle and FilmLoop were setting up shop downtown, as close to Stanford as possible for recruiting purposes, recapturing the abandoned real estate. That's where his peers in the program chose to go. But Palo Alto was a boring place to spend a summer.

Systrom read in the *New York Times* about a trend in online audio, and saw mention of a company called Odeo, which made a marketplace for podcasts on the internet. That's where he decided he wanted to intern. He sent a chance email to CEO Evan Williams, who was a couple years into working on the startup, which was based a 45-minute drive north, in San Francisco. Williams was already tech-world famous for selling Blogger, a blogging website, to Google. Systrom got the internship. He took the train every day into the city, which was more exciting, with its quality whiskey bars and live music scene.

• • • • • • • • • • • •

Jack Dorsey, a new engineering hire at Odeo, was expecting to dislike the 22-year-old intern he had to sit next to all summer. He imagined that an exclusive entrepreneurship program and an elite East Coast boarding school were both sterile, formulaic places, and that a person shaped by them might be devoid of creativity.

Dorsey, a 29-year-old New York University dropout with an anarchist tattoo and a nose ring, considered himself to be more of an artist. He would sometimes dream, for instance, about becoming a dressmaker. He was an engineer, but only as a means to an end—to create something out of nothing, with code. Also, so he could pay rent. He was not the kind of person who knew what to do with an intern.

To Dorsey's surprise, he and Systrom became fast friends. There were only a handful of employees in the loft on Brannan Street, most of them vegan, so he and Systrom bonded over lunchtime walks for sandwiches from the local deli. It turned out that they both had very specific tastes in music and an appreciation for high-quality coffee. They both liked photography. There weren't many engineers in Silicon Valley Dorsey could talk to about those things. And Systrom flattered Dorsey, who was self-taught, by asking for his help with computer programming.

Systrom did have his quirks. Once he learned to get better at the coding language JavaScript, he was precious about perfecting its syntax and style so that it was nice to look at. This made no sense to Dorsey and was almost sacrilegious in Silicon Valley's hacker culture, which revered getting things done quickly. It didn't matter if you sealed lines of text together with the digital equivalent of duct tape, as long as it worked. Nobody cared about the code having beautiful structure, except Systrom.

Systrom would wax philosophical about his other highbrow interests that Dorsey had never had the opportunity to develop. Still, Dorsey saw a bit of himself in the intern, who seemed to know enough about culture to have opinions about it, and was not simply trying to be a cog in a machine or get rich, like the others with a business education. Dorsey was curious about what would come for Systrom, once he relaxed a bit. But he later found out that Systrom, after graduation, was planning to take a job at Google. In product marketing. *Figures,* Dorsey thought. He was a typical Stanford guy after all.

• • • • • • • • • • • • •

His last year at Stanford, in between Odeo and Google, Systrom worked to make some side money pulling espresso shots at Caffé del Doge on University Avenue in Palo Alto. One day, Zuckerberg walked in, puzzled to find the student he'd tried to recruit working at a coffee shop. Even back then, the CEO wasn't comfortable with rejection. He awkwardly ordered and moved on.

TheFacebook.com, now simply called Facebook, ended up launching photos in October 2005, without Systrom's help. The added invention, two months later, of tagging friends in photos proved even more fruitful for the company. People who weren't yet using Facebook were suddenly getting email alerts that photos with their faces in them were appearing on the website, and were tempted to click to see. It became one of Facebook's most important manipulations for getting more people to use the social network, despite the hint of creepiness.

Systrom felt a tinge of missed opportunity. More than 5 million people were using Facebook now, and he realized he must have been wrong about its trajectory. He tried to go back, and started reaching out to one of the employees running product under Zuckerberg. But the person stopped answering his emails, which he assumed meant they weren't interested.

The team at Odeo was launching a new status update product, called Twttr, pronounced "twitter," with Dorsey as its CEO. Systrom had kept in touch and used the site frequently to support his friends and former colleagues, posting about whatever he was cooking or drinking or looking at, even though the site was text only. One of the guys at Odeo told him that eventually, celebrities and brands around the world were going to use it to communicate. *They're crazy,* Systrom thought. *Nobody is going to use this thing.* He couldn't imagine what it might be useful for. Either way, they didn't try to recruit him back.

Most people never get the chance to join an iconic company in its early days. Systrom squandered both of his, choosing instead to do something much less risky. For him, after graduating Stanford with degrees in management and engineering, going to Google was basically like going to grad school. He'd have a salary with a base of about $60,000—paltry compared to the life-changing wealth Facebook would have afforded—but he would get a crash course in Silicon Valley logic.

Founded in 1998, Google had started trading on public markets in 2004, minting enough millionaires to lift Silicon Valley out of its malaise from the dot-com bust. When Systrom joined in 2006, it had almost 10,000 employees. Google, far more functional and established than tiny Odeo, was led mostly by former Stanford students making data-based decisions. It was the culture that drove homepage leader Marissa Mayer, who later became CEO of Yahoo!, to famously test 41 shades of blue to figure out what color would give the company's hyperlinks the highest click-through rate. A slightly purpler blue shade won out over slightly greener shades, helping boost revenue by $200 million a year. Seemingly insignificant changes could make a huge difference when applied to millions or billions of people.

The search company did thousands of tests like this—known as A/B tests, showing a different product experience to different segments of the user base. At Google, every problem was imagined to have a right answer, which could be determined through quantitative analysis. The company's methods reminded Systrom of the prodigy kids in his computer science class, trying to do something overly complicated to impress. It was easy, in those instances, to solve the wrong problem. If Googlers were studying photography, for example, they might have aimed to make the best camera, instead of the most striking photograph. Charlie would have been alarmed.

More exciting, Systrom thought, was when Google employees would break out of the definitive methods and use their intuition. He worked writing marketing copy for Gmail, where the team was trying to figure out how to get users their email faster. Their solution was creative: as soon as a person went to Gmail.com and started entering her username, Google would start downloading the data for her inbox while she typed in her password. Once she clicked to log in, some of her emails would be ready to read, leading to a better user experience without requiring a faster internet connection.

Google was not interested in letting Systrom make products, since he didn't have a computer science degree. He was so bored with writing marketing copy that he started teaching a younger colleague to make latte art with the corporate espresso machines. Eventually he switched over to Google's deals team, watching how the technology giant courted and then acquired smaller companies. He would make PowerPoint presentations analyzing targets and marketing opportunities. There was only one problem: in 2008, the U.S. economy fell into crisis from mortgage defaults. Google wasn't buying anything.

"What should I do?" Systrom asked one of his colleagues.

"You should pick up golf," the colleague suggested.

I'm too young to pick up golf, Systrom decided. It was time to move on.

• • • • • • • • • • • •

By age 25 Systrom had received an introduction to how growth-driven Facebook was, how scrappy Twitter was, and how procedural and academic Google was. He was able to know their leaders and understand what drove them, which stripped them of their mystery. From the outside, Silicon Valley looked like it was run by geniuses. From the inside, it was clear that everyone was vulnerable, like he was, just figuring it out as they went along. Systrom wasn't a nerd, or a hacker, or a quant. But he was perhaps no less qualified to be an entrepreneur.

Still too risk-averse to start something without a salary, he took a job as a product manager at a tiny startup called Nextstop that made a website for people to share their travel tips. Meanwhile, on nights and weekends in cafes, he tried to build a new skill: making mobile apps.

San Francisco's coffee shops in 2009 were full of people like Systrom, tinkering on the side, betting that mobile phones would usher in the next technology gold rush, with a much bigger opportunity than Web 2.0. After Apple introduced the iPhone in 2007, smartphones started to change the way people thought about going online. The internet was no longer just for accomplishing tasks, like checking email or searching on Google—it was now something that could be enmeshed with regular life, as people carried it in their pockets.

Developers could now offer completely novel types of software that could go wherever people went. Big web services like Facebook and Pandora were among the most popular apps in spring 2009, but so were gimmicky tools like Bikini Blast, which offered racy background images for your phone, and iFart, an app that made various flatulence noises depending which button you pushed. The apps race was a free-for-all, led by mostly male twenty-somethings in San Francisco throwing ideas at the public to see what would stick.

Systrom thought that what he lacked in technical ability—he didn't actually know how to make an app, only a mobile website—he could make up for with his relative well-roundedness, which he hoped would help him come up with ideas that were more fun and interesting for regular people. He was learning to develop just through practicing, the same way he'd learned to DJ, or to make a leaf pattern in latte foam, or to become a better photographer. He made a handful of random tools, like a service called Dishd for people to rate meals instead of restaurants. His Stanford friend Gregor Hochmuth helped him on it, building a tool to crawl the web for restaurant menus, so a user could search for an ingredient like "tuna" and find all the places that served it.

Later that year, Systrom built something called Burbn, after the Kentucky whiskey he enjoyed drinking. The mobile website was perfect for Systrom's urban social life. It let people say where they were, or where they planned to go so their friends could show up. The more times a user went out, the more virtual prizes they got. The background color scheme was an unattractive brown and red, like a bottle of bourbon with a red wax topper. In order to add a picture to your post, you had to email it in. There was no other technical way to do it. Still, it was an idea good enough to enter the Silicon Valley apps race.

In January 2010, determined to make his pitch and justify quitting Nextstop, Systrom headed to a party for a startup called Hunch at Madrone Art Bar in San Francisco's Panhandle neighborhood. It would be swarming with venture capitalists, mostly because of Hunch's already-successful executives: Caterina Fake, a cofounder of Flickr, a photo storage and sharing website that had sold to Yahoo! for a reported $35 million in 2005, and Chris Dixon, who'd sold a security company he cofounded in 2006.

Over cocktails, Systrom met two important VCs with checkbooks: Marc Andreessen, a cofounder of Netscape, who ran Andreessen Horowitz, one of the valley's hottest venture capital firms, and Steve Anderson, who ran a much quieter early-stage investing shop called Baseline Ventures.

Anderson liked the fact that Systrom, with pedigrees from Stanford and Google and a confident personality, didn't have any investors yet for his mobile idea. Anderson liked to be the first to notice something. He borrowed Systrom's phone to type an email to himself: "Follow up."

From there, the two of them met every couple weeks at the Grove on Chestnut Street, ordering cappuccinos and talking about Burbn's potential. Systrom's program had just a few dozen people using it—his friends and their friends. He said he needed about $50,000 to get started making it a real company. Anderson was interested in the opportunity, but on one condition.

"The biggest risk for you is you're a sole founder," Anderson told Systrom. "I usually don't invest in sole founders." He argued that without someone else at the top, nobody would tell Systrom when he was wrong, or push his ideas to be better.

Systrom said he agreed, and would carve out a 10 percent equity in the term sheet for an eventual cofounder. Just like that, the company that would become Instagram got its start.

• • • • • • • • • • • •

Hochmuth, Systrom's app-tinkering buddy, was the obvious person to build a company with. But he was happy at Google. "Why don't you talk to Mikey?" Hochmuth suggested.

Mike Krieger was another Stanford student, two grades younger, whom Systrom knew from the Mayfield fellowship. Systrom had first met Krieger years earlier at a Mayfield networking event, where Krieger read Systrom's Odeo name badge and quizzed him about what that company was like. Then Krieger disappeared for a while to complete a master's degree in "symbolic systems"—the famous Stanford program for understanding the psychology of how humans interact with computers. He wrote his thesis about Wikipedia, which had somehow cultivated a community of volunteers to update and edit its online encyclopedia. In 2010 he was working at Meebo, an instant-messaging service.

Systrom liked Krieger quite a bit. He was good-natured, levelheaded, always smiling, and a much more experienced engineer than himself. Krieger had straight brown hair just long enough to be floppy, with a clean-shaven oval face and rectangular glasses. Recently, Systrom and Krieger had been running into each other at a San Francisco cafe called Coffee Bar on the weekends, where they were giving each other feedback on side projects and exchanging advice. Krieger was one of the early Burbn testers and liked it because it included visual media, not just status updates.

Krieger, like Systrom, had had no idea he'd end up in the startup world. He grew up in Brazil, with occasional stints in Portugal and Argentina, since his father worked for the beverage company Seagram. He also enjoyed music, and could play a 12-string guitar. He'd dabbled in website design in high school, but he had never met any technology entrepreneurs. After arriving in the United States in 2004 for Stanford, he'd quickly realized the industry could be a fit.

Krieger's plan was to start at Meebo, a medium-size company, then graduate to a smaller, more challenging one, then eventually start his own a few years later once he knew enough. In the meantime, he was dabbling with iPhone app development at cafes. The first one he built, with the help of a talented designer friend, was called Crime Desk SF. It overlaid San Francisco crime data from public records on top of a camera tool to look out at the real world, seeing what had occurred nearby. They had spent too much time making it beautiful. Unfortunately, nobody wanted to use it.

Krieger had told Systrom that he'd be willing to help out, if Systrom ever needed another hand with Burbn. After the investment from Anderson, Systrom told Krieger the idea was turning into a real company, with real financial responsibilities, and asked Krieger if he wanted to be an official cofounder.

"Count me interested," Krieger said. It seemed obvious to him: he could work in San Francisco instead of commuting south to Meebo in the Valley, he could help build something in the cool new mobile app space, and he could do it with this guy he already liked talking to.

Krieger often had strong gut feelings when making important decisions. But he would always try to be strategic about how to get others on his side. In this case, he knew his mom and dad in São Paulo would be worried about their son making such an impulsive career choice, on an immigrant visa. So he presented the idea to them in steps.

"Hey, I'm thinking it would be interesting to join a brand-new startup!" he told them in Portuguese, framing it as something he might do eventually, if he found the right opportunity.

A few days later, he called again.

"Hey, I met this interesting guy!" He explained who Systrom was and what he was working on.

By the end of the week, he finally called to tell them that after all his research, he was deciding to be a cofounder in Systrom's company, Burbn. His parents, under the impression that their son had taken his time to make the choice, were supportive.

● ● ● ● ● ● ● ● ● ● ● ●

The next entity to convince was the United States government. That January 2010, Krieger hired an immigration lawyer with experience working on Brazilian visas (though most of her prior clients were hairdressers). He applied to switch his immigrant work visa over to Burbn. The government officials reviewing his case saw that Burbn had raised money but were suspicious—was there a business plan?

Of course there wasn't. Their funding would allow them to do the same thing as Facebook: try to make their product part of a daily habit for its users before trying to make money off them. But Krieger and Systrom couldn't say that. They told the government they were one day planning to make money off a kind of local coupon system, for the bars, restaurants, and stores people told their friends they were at. They explained that their competitors included Foursquare and Gowalla. They also provided a chart, which predicted that by their third year, they would likely have 1 million users. They laughed about how improbable that was.

As they were waiting to hear whether it was legal to work on Burbn together, Krieger and Systrom tried to test whether they actually liked working together, period. They spent a few weeknights at Farley's, a Potrero Hill coffee shop that displayed the work of local artists on the wall. They coded little games that would never be released, including one based on the prisoner's dilemma, a political game theory that explains why rational people might not cooperate even when they should.

It was fun, but it wasn't Burbn. Months went by, and Krieger understood that Systrom was spending his cash, delaying his progress, without an obvious end to the wait. Krieger was spending hours reading up on immigration law and obsessing over horror stories that people posted on internet forums.

"Kev, maybe you should pick a different cofounder," Krieger would suggest.

"No, I really want to work with you," Systrom would respond. "We'll figure this thing out."

Systrom had seen enough startups with toxic cofounder relationships to know how rare it was to find someone he could trust. The founders at Twitter, for example, were always trying to undermine one another. Dorsey was actually no longer the CEO of the company. Employees complained he had been taking credit for all the ideas and success around Twitter while avoiding managing people. Dorsey would take breaks for hot yoga and sewing classes. "You can either be a dressmaker or the CEO of Twitter," Ev Williams said to him, according to Nick Bilton's book *Hatching Twitter*. "But you can't be both." In 2008, Williams worked with Twitter's board to take over, ousting Dorsey.

Facebook's story was even more dramatic. Cofounder Eduardo Saverin, who started to feel left out of company decisions when the team moved to Palo Alto in 2005, froze the Facebook bank account—which may have been the real reason Zuckerberg's card didn't work during his first meal with Systrom. Zuckerberg's lawyers devised a complicated financial transaction to dilute Saverin's ownership stake, spurring a lawsuit and a dramatic Hollywood adaptation of the story, the movie *The Social Network*, which would come out later in 2010.

Founders of Silicon Valley lore were aggressive, ambitious, controlling, emotionless. Krieger was a good listener, an attentive partner, a hard worker, and, after all their test runs together, a good friend. Systrom wasn't going to risk partnering with anyone else.

• • • • • • • • • • • •

All the while, Systrom was trying to find more backers for their project. He managed to convince Andreessen Horowitz to put in $250,000, through a connection to Ronny Conway, a partner at the firm he knew from Google. Once Baseline's Anderson heard that number, he wanted the same level of ownership, so he bumped his investment to $250,000 too. Suddenly, Systrom had half a million dollars to work with.

Anderson tried to drum up other interest for Burbn, emailing about a dozen peers at other firms, but anyone who hadn't been personally charmed by Systrom was uninterested. There were several more popular location-based apps, like Foursquare and Gowalla, and photos weren't enough of a killer feature to bring people in, VCs told him. Burbn had social qualities, but Facebook was already so dominant in that area, it didn't make sense to bet against them. Status updates—about what you were doing or where you were going—were already big at Twitter.

So Systrom reached out to his old mentor Dorsey, letting him know that he was starting a company. The two of them met near Dorsey's offices for Square, his latest entrepreneurial adventure. Dorsey was creating a piece of hardware you could plug into your computer or phone, linked to the internet, that would allow people to buy things with credit cards anywhere. The nose ring was gone. Dorsey was dressing much more formally, with crisp white designer dress shirts by Dior and a black blazer, perhaps in reaction to the distrust from Twitter's board.

Dorsey asked Systrom a lot of the same questions the VCs had, about why anyone would use Burbn instead of Foursquare. *Of course he named it after bourbon*, Dorsey thought, remembering Systrom's highbrow interests. *Of course he used the trendiest coding language*. Systrom, who was still learning how to make regular iPhone apps, sold the idea that an app built for the mobile web in HTML5 would have an advantage in the market, which Dorsey wasn't sure about. But in this case, personal relationships trumped investment logic. Honestly, Dorsey thought, it didn't matter what Systrom was building. There were no benchmarks or models that dictated what would and wouldn't win in mobile, anyway. And Systrom had asked him at exactly the right time.

Nobody had ever thought to ask Dorsey to put his money in a startup. If he did the Burbn deal, it would be his first "angel investment," which is what a small, early startup investment from a rich person was termed in the Valley. It would be a cool thing to do with his newfound wealth from Twitter, while supporting Systrom, who he thought had exceedingly good taste. Systrom would figure out Burbn, whatever it became.

Dorsey lent his support, to the tune of $25,000. His cheerleading would turn out to be far more valuable than that.

• • • • • • • • • • • • •

The government finally approved Krieger's visa in April 2010, almost three months after he applied. His first week at the new company, Systrom took him to breakfast and made a confession: he wasn't sure Burbn was the right product to be building.

Systrom explained that the idea resonated with their young hipster friends in cities, who were going to music shows and restaurants. Burbn's prizes for being social were fun because they made the experience addictive and competitive. But anyone who wasn't a young urbanite—a parent, for example, or someone without money to go out—might not *need* to use it. Even Dorsey would log on only after Systrom asked him for feedback. Systrom thought of how scary it must have been for the team at Odeo to switch to building Twitter, but clearly that had been the right decision. What was their Twitter?

Krieger was caught off guard. He had just taken a major risk to come work for Systrom on Burbn, not just by leaving a more secure job. If they started a new company and ran out of money working on it, he'd be back to the visa process, or back to Brazil. Before throwing it all away, Krieger argued, perhaps they should try to improve it. So they did, building an iPhone app version.

The cofounders graduated out of their meetings at local coffee shops and into a rickety coworking space called Dogpatch Labs, on a pier near San Francisco's ballpark, where the other small startups included Threadsy, TaskRabbit, and Automattic, the maker of WordPress.

It was a strange, drafty place, producing a cacophony of distracting sounds: screeching seagulls and barking sea lions, but mostly the sound of other young people being creative and sometimes unproductive, emboldened by Red Bull and alcohol. On the ceiling, an enormous ship wheel hovered in a display of nautical kitsch but also danger, as it could fall in an earthquake. The surrounding water was cold. Few tourists were brave enough to rent the kayaks at the outside stand. But on Friday afternoons, when engineers would gather outside for happy hour, someone would inevitably get too drunk and decide to jump into the San Francisco Bay.

Krieger and Systrom kept typing, trying to ignore their peers, wondering if everyone else was less worried than they were about running out of money. The Burbn founders took advantage of the social events another way. The building manager told them that if anyone ever got catered food, they could take the leftovers for free after 1:30 p.m. If they got hungry before then, they'd buy the sandwich on special for $3.40 at the local bodega.

They had to save money because they weren't sure how long it would take to make Burbn successful, or if it would even be successful. A couple months later, a meeting with Andreessen's Conway, the son of famed Silicon Valley angel investor Ron Conway, dashed their hopes further.

"What do you guys do again?" Conway asked. Systrom tried once more to explain Burbn—*It's a fun way to see what your friends are doing and go join them in real life! You can get inspired about where to go next!* But it was clear Conway wasn't excited about the idea, despite playing a part in his firm's investment. To him, Systrom seemed to be rattling off all of Silicon Valley's latest buzzwords. Mobile? *Check*. Social? *Check*. Location-based? *Check*.

Conway was probably the tenth person with a deer-in-the-headlights reaction to the app, Systrom thought. *He has no interest or faith in what we're working on, despite their investment*. Systrom knew their product was fun, but was it *useful*? Did it solve a problem most people had in their lives? The question was a tipping point, sending Systrom and Krieger back to the drawing board.

. . . .

The founders took over a whiteboard in one of the Dogpatch Labs conference rooms and had a brainstorming session that would serve as the foundation for their entire leadership philosophy: to ask first what problem they were solving, and then to try and solve it in the simplest way possible.

Krieger and Systrom started the exercise by making a list of the top three things people liked about Burbn. One was Plans, the feature where people could say where they were going so friends could join them. Another was photos. The third was a tool to win meaningless virtual prizes for your activity, which was mostly a gimmick to get people to log back in.

Not everybody needed plans or prizes. Systrom circled "photos." Photos, they decided, were ubiquitous, useful to everybody, not just young city dwellers.

"There's something around photos," Kevin said. His iPhone 3G took terrible pictures, but it was only the beginning of that technology. "I think there will be an inflection point where people don't carry around point-and-shoots anymore, they're just going to carry around these phones."

Everyone with a smartphone would be an amateur photographer, if they wanted to be.

So if photos were the killer feature of the app they should build, what were the main opportunities? On the whiteboard, Systrom and Krieger brainstormed three of the top problems to solve. One, images always took forever to load on 3G cellular networks. Two, people were often embarrassed to share their low-quality phone snaps, since phones weren't nearly as good as digital cameras. Three, it was annoying to have to post photos in many different places. What if they made a social network that came with an option to deliver your photos to Foursquare, Facebook, Twitter, and Tumblr all at once? Playing nice with the new social giants would be easier than competing with them. Instead of having to build a network from scratch, the app could just piggyback off already-established communities.

"All right," Systrom said. "Let's focus on photos, and on solving these three problems." They would make it an app for iPhone only, since Krieger was better at those. Systrom's argument to Dorsey, that the trendy HTML5 coding language would be a helpful differentiator in the marketplace, turned out to be wrong. They would have to make the app useful first, and add Android later, if they were lucky enough to become that popular.

* * * * * * * * * * * * *

Their first prototype was named Scotch, a relative to bourbon. It allowed people to swipe through photos horizontally and tap to like them, similar to a Tinder before its time. They used it for a few days before going back to the Burbn idea, doubting their instincts. And then they tried a new concept that would allow people to scroll through photos vertically, showing the most recent post first, like Twitter.

All of the photos would use as few pixels as possible, so that they would load quickly, helping solve problem number one—only 306 pixels across, the minimum required to display a photo on an iPhone with 7-pixel borders on each side. The photos would be square, giving users the same creative constraint for photography as Systrom's teacher in Florence gave him. It was similar to how Twitter only let people tweet in 140-character bursts. That would help solve, but not fully solve, problem number two.

There were two different kinds of social networks one could build—the Facebook kind, where people become mutual friends with each other, or the Twitter kind, where people follow others they don't necessarily know. They thought the latter would be more fun for photos, because then people could follow based on interests, not just friendship.

Displaying "Followers" and "Following" at the top of the app, the way Twitter did, made it just competitive enough that people would need to come back to the app and check their progress. People could also "like" something, appending a heart, similar to Facebook's thumbs-up. Liking was much easier on this new app, because you could do it by double tapping on an entire photo instead of looking for a small button to click. And unlike on Twitter and Facebook, nobody on this new app needed to come up with anything clever to say. They simply had to post a photo of what they were seeing around them.

If Systrom and Krieger wanted to fully copy Twitter's concepts, it would be obvious, at this point, to add a reshare button, to help content go viral like the retweet did. But the founders hesitated. If what people were sharing on this app was photography, would it make sense to allow them to share other people's art and experiences under their own names? Maybe. But in the interest of starting simple, they decided not to think about it until post-launch.

They picked a logo—a version of a white Polaroid camera. But what to call it? The vowel-less alcohol theme was getting to be too cute. Something like "Whsky" wouldn't necessarily explain what the app was for. So they tabled the discussion, calling it Codename.

Soon after, Systrom and the girlfriend who would become his wife, Nicole Schuetz, whom he'd met at Stanford, went on a short vacation to a village in Baja California Sur, Mexico, called Todos Santos, with picturesque white sand beaches and cobblestone streets. During one of their ocean walks, she warned him that she probably wouldn't be using his new app. None of her smartphone photos were ever good—not as good as their friend Hochmuth's were, at least.

"You know what he does to those photos, right?" Systrom said.

"He just takes good photos," she said.

"No, no, he puts them through filter apps," Systrom explained. Phone cameras produced blurry images that were badly lit. It was like everyone who was buying a smartphone was getting the digital equivalent of the tiny plastic camera Systrom used in Florence. The filter apps allowed users to take an approach similar to that of Systrom's professor, altering photos after they were captured to make them look more artsy. You didn't have to actually be a good photographer. Hipstamatic, with which you could make your photos look oversaturated, blurred, or hipster vintage, would be named Apple's app of the year in 2010. Camera+, another editing app, was another one of the most popular.

"Well, you guys should probably have filters too," Schuetz said.

Systrom realized she was right. If people were going to filter their photos anyway, might as well have them do it right within the app, competition be damned.

Back at the hotel, he researched online about how to code filters. He played around on Photoshop to create the style he wanted—some heavy shadow and contrast, as well as some shading around the edges of the image for a vignette effect. Then, sitting on one of the outdoor lounge chairs with a beer beside him and his laptop open, he set about writing it into reality.

He called the filter X-Pro II, a nod to the analog photo development technique called cross-processing, in which photographers intentionally use a chemical meant for a different type of film.

Soon after, he tested his work on a photo he took of a sandy-colored dog he came across in front of a taco stand. The dog is looking up at Schuetz, whose sandaled foot appears in the corner of the shot. And that, on July 16, 2010, was the first-ever photo posted on the app that would become Instagram.

• • • • • • • • • • • •

Krieger and Systrom had no idea whether their new app would appeal to anyone any more than Burbn did. Nothing about it was new, exactly. They were not the first to think of photo filters or interest-based social networks. But the founders valued feel and simplicity over technological innovation. By keeping the product minimalist—just for posting and liking photos—they would spend less time developing it, and would be able to test it on the public before spending any more money. They set a deadline to launch whatever Codename would be in eight weeks, less time than it had taken Krieger to get his visa.

While they were building it, they received an unsolicited email from Cole Rise—a local designer who had heard what they were working on and wanted to be a tester for it.

Rise was the perfect candidate. He was working at a video startup and was also a photographer. The theme for his photography, which he occasionally displayed in local galleries, went against the market trend of crisp, perfect, higher-resolution images. He digitally manipulated his photos to let more light leak in, or added more texture or feeling to make them more nostalgic. He appreciated vintage cameras, like Polaroids, and had just purchased a Hasselblad, a variant of the camera used in the first moon landing. It only took pictures in square format.

After Systrom and Krieger agreed to let Rise test the app, he took his phone hiking at Mount Tamalpais, north of the city. He tried one of Systrom's filters, called Earlybird, and was floored by the quality, thinking it similar to that of his own art. He asked the founders out for drinks.

They met at Smuggler's Cove, a shipwreck-themed rum bar in San Francisco that served premium flaming cocktail bowls. Systrom and Krieger asked Rise a lot of questions about his beta-testing experience, and he started to sense that the founders didn't know their potential.

"This is going to be fucking huge," Rise explained. In the tech industry, leaders rarely had any experience in the industry they were disrupting. Amazon's Jeff Bezos had never been in books and Tesla's Elon Musk had never been in car manufacturing, but Instagram's filters had clearly been made by a photographer. Earlybird was the best Rise had ever seen, he explained—far higher quality than anything on Hipstamatic.

After a few drinks, the founders asked Rise if he would like to create some filters of his own, as a contract job. Rise agreed, thinking it would save time to have an app that would automatically edit his pictures exactly how he wanted them to be edited. He'd built up a complicated system after spending years collecting textures from things he saw around him. He would overlay those textures on files in Adobe Photoshop, then add layers of color change and curves.

Rise tested each of his ideas on twenty different images from his camera roll, of sunrises, sunsets, different colors and different times of day. He ended up turning in four filters, which were called Amaro, Hudson, Sutro, and Spectra. He didn't think about the long-term consequences of giving up his art to a company, making it available to the masses. Optimistic though he was for his new friends, he knew that most startups failed.

Neither Rise nor the founders thought there was a downside to the fact that filters, when used en masse, would give Instagrammers permission to present their reality as more interesting and beautiful than it actually was. That was exactly what would help make the product popular. Instagram posts would be art, and art was a form of commentary on life. The app would give people the gift of expression, but also escapism.

• • • • • • • • • • • • •

Late one night, lit by the glow of his laptop in rickety Dogpatch Labs, Systrom was coding in a corner, trying not to be distracted by the fact that there was an entrepreneur pitch event going on. A man named Travis Kalanick was in front of an audience of mostly men explaining his company, UberCab, which made a tool that was supposed to help people summon luxury cars with their phones. It would officially launch in San Francisco the next year.

One of the event's guests was Lowercase Capital's Chris Sacca, an early investor in Twitter, who was already putting money in UberCab. Sacca considered himself a good judge of character, and had made a call to invest in Kalanick after inviting him for hours of hot-tubbing at his Lake Tahoe home. He recognized Systrom in the corner. They had overlapped at Google, briefly, before Sacca left to found Lowercase Capital. If Systrom was here, coding at night, he must be working on something new, Sacca thought.

After Sacca approached Systrom, he was invited to learn more about the product at Compass Cafe, a nearby coffee shop run by former inmates transitioning back into society. There, Systrom showed off the latest version of Codename.

"How big can photos really be?" Sacca inquired. In venture capital, investors take bigger risks because they expect many multiples in return on their investment. Sacca had already been an investor in Photobucket, which had sold to News Corp's Fox Interactive Media for $330 million, and had watched Flickr sell to Yahoo! for $35 million. If this was another take on Twitter, he'd seen dozens of attempts to do the same, all of which fizzled.

Systrom didn't make any predictions. Instead he leaned on his Stanford business education and attempted to sell exclusivity. "I'm only inviting three angel investors into the deal," he said. "It's you, Jack Dorsey, and Adam D'Angelo." D'Angelo was the founder of Quora and previously the chief technology officer of Facebook, whom Systrom had met when he was a Stanford student.

The flattery worked. "That's pretty badass," Sacca said. Then he asked about some features he sensed were missing.

"When we get to ten, fifty million users, we might be able to turn that on, but for now we're just focused on keeping the product simple," Systrom answered.

Sacca was floored. Millions of users? Systrom had fewer than 100 beta testers. Sacca received so many pitches from entrepreneurs who talked up the bells and whistles of their products with polished presentations, but here Systrom was just calmly assuming success was inevitable, asking whether Sacca wanted in. He did.

•   •   •   •   •   •   •   •   •   •   •   •   •

Systrom and Krieger wanted to come up with a name that was easy to pronounce—and spell, after Burbn. They also wanted it to portray a sense of speed in communication. They'd borrowed Gmail's trick, and would start uploading photos while users were still deciding which filter to apply. A lot of the good photo-related startup names were taken, so they came up with "Instagram," a combo of "instant" and "telegram."

They could already see that their decision to share images to Facebook and Twitter had a powerful side effect. Every time users chose to share one of their Instagram photos somewhere else, new people using other social media sites would see it and potentially check out Instagram and download the app.

The founders picked their first users carefully, courting people who would be good photographers—especially designers who had high Twitter follower counts. Those first users would help set the right artistic tone, creating good content for everyone else to look at, in what was essentially the first-ever Instagram influencer campaign, years before that would become a concept.

Dorsey became their best salesman. He was initially shocked to find out his investment money was going toward an entirely different app than Burbn. Usually, founders pivoted to a new product as a last-ditch effort to avoid going out of business. But Dorsey loved Instagram, way more than he'd ever loved Burbn.

With Dorsey's first photo, of a baseball game from a tech investor's box at the Giants' stadium, he was amazed to find the filter immediately made the field look greener. He'd just gotten his first car and wanted to use it, so on the weekends he'd drive thirty minutes south to the Ritz-Carlton at Half Moon Bay to sit and read the paper by outdoor fire pits. He would take a lot of pictures for Instagram along the way.

Once Dorsey was addicted to Instagram, the product seemed so obvious and useful that he wished Twitter had managed to build it first. He asked Systrom if he would be open to Twitter acquiring his company. Systrom sounded enthusiastic.

But Dorsey had spoken too soon. When he emailed Williams, telling him about the idea, the rejection was loaded with the bitterness Williams felt toward Dorsey personally. Williams was CEO and was still trying to establish himself as Twitter's leader. Dorsey's strategy was not welcome.

"We already looked at that," Williams replied. It was true. Systrom had already reached out to Williams and tried to meet with him. Twitter didn't know if he was looking for a buyer, but the deals team had done its research anyway, and postulated that they could buy Instagram for about $20 million. But Williams wasn't excited about the product. He thought that Instagram would be for frivolous posts, for people taking artsy pictures of their lattes, like Dorsey on his Half Moon Bay jaunts. It was not for anything that would be in the news, and not for any of the serious world-changing conversations Twitter was about. "We don't think it's going to be very big," he told Dorsey.

After that, Dorsey had another motivation to promote Instagram—to prove Williams wrong. Everything he posted on Instagram would immediately cross-post to Twitter, reaching his 1.6 million followers there. He told the world it was his favorite new iPhone app, and they listened.

. . . .

When Instagram launched to the public on October 6, 2010, it immediately went viral thanks to shares from people like Dorsey. It reached number one in camera applications in the Apple app store.

Instagram only had a single computer server processing all the activity remotely, at a data center in Los Angeles. So Systrom was in a state of panic, wondering if everything was going to fall apart, and whether everyone was going to think he and Krieger were idiots.

Krieger nodded and smiled, not indulging Systrom's fear, mostly because it was unproductive. With the unexpected deluge of users, they needed to think fast to keep Instagram online.

Systrom called D'Angelo, the early Facebook chief technology officer and early investor in Instagram, for his advice. It was the first of several calls that day. Every hour, Instagram seemed to grow faster. D'Angelo eventually helped the company transition to renting server space from Amazon Web Services instead of buying their own.

Within the first day, 25,000 people were using Instagram. Within the first week, it was 100,000, and Systrom had the surreal experience of seeing a stranger scrolling through the app on a San Francisco bus. He and Krieger started an Excel spreadsheet that would update live with each user added.

Launch success is rarely a sign of an app's longevity. People download new apps, get excited, and forget to open them again. But Instagram remained a sensation. Around the holiday season, Krieger and Systrom took a break from the infrastructure scares to huddle around the computer screen, Belgian beers in hand, and watch the number in the Excel spreadsheet tick up to 1 million. Six weeks after that, they were at 2 million.

• • • • • • • • • • • •

Articles would later reflect on Instagram's origin, crediting the app with perfect timing. It was born in Silicon Valley, in the midst of a mobile revolution, in which millions of new smartphone consumers didn't understand what to do with a camera in their pockets. That much is true. But Systrom and Krieger also made a lot of counterintuitive choices to set Instagram apart.

Instead of continuing to build the app they'd originally promised investors, the cofounders stopped and tried a bigger idea. They aimed to do just one thing—photography—really well. In that sense, their story is similar to Odeo's, when Dorsey and Williams switched gears to focus on Twitter.

Instead of trying to get everyone to use their app, they invited only people they thought would be likely to spread the word to their followers elsewhere, especially designers and creatives. They sold exclusivity to investors, even when so many of them were skeptical. In that sense, they were like a luxury brand, manufacturing coolness and tastefulness around what they'd built.

And instead of inventing something new and bold, as potential Silicon Valley investors wanted them to, they improved on what they'd seen other apps do. They made a tool that was much simpler and faster to use than anyone else's, taking up less of users' time as they were out living the experiences Instagram wanted them to capture. And they had to do it via their phones, because there was no Instagram website, making those experiences feel immediate and intimate to others.

Instagram's simplicity helped it catch on the way early Facebook did, when Zuckerberg reacted to the loudness of Myspace with a clean design. By the time Instagram launched, Facebook was crowded with features—it had the news feed, events, groups, and even virtual credits to buy birthday gifts—and was already plagued with privacy scandals. On Facebook, posting a photo from a mobile phone was a hassle. All photos had to be uploaded as part of a Facebook Album, a tool designed for people with digital cameras. Any time someone added a photo from their phone to Facebook, it would join a default album called Mobile Uploads. That created an opening for Instagram.

Beyond the product's mechanics, the founders excelled in playing off the strengths of other people and other companies. They realized they weren't starting from scratch. The technology industry already had winners; if Instagram made the giants look good, they'd get a boost in the process. Instagram was a crown jewel of Apple's app store, later featured onstage at iPhone launches. It became one of the first startups to thrive on Amazon's cloud computing. It was the easiest way to share photos on Twitter.

The upside of this collaborative strategy was that one day, Instagram would get to be a giant too. But the founders would have to make many painful compromises along the way.

· · · · · · · · · · · · ·

Back in December 2010, two months after Instagram's launch, Systrom was home in Holliston, Massachusetts, for Christmas. Dennis Crowley, the Foursquare CEO, had grown up in Medway, the bordering town, of the same suburban quality with trees and creeks, and Systrom reached out to see if he would meet. No longer competing, they met up for drinks at Medway Lotus, a Chinese restaurant slash karaoke bar.

Systrom was now fielding constant outreach from people who wanted to invest, as well as from representatives from large companies like Google and Facebook, offering help and advice—which Systrom understood as grooming for acquisition interest.

He explained to Crowley that everything was starting to click. He understood the opportunity now. Everyone took photos on their phones, and everyone wanted them to look better. Everyone was going to use Instagram.

"One day, Instagram is going to be bigger than Twitter," he predicted, feeling bold.

"There's no way!" Crowley pushed back. "You're crazy."

"Think about it," Systrom urged. "It's so much work to tweet. There's a lot of pressure about what you're going to say. But it's so easy to post a photo."

Crowley thought about it. But, he argued, so many other photo technology services had come and gone without changing the world. What made Instagram any different than those?

Systrom didn't have an insightful answer, except to note that it seemed to be catching on. Instagram's early popularity was less about the technology and more about the psychology—about how it made people feel. The filters made reality look like art. And then, in cataloging that art, people would start to think about their lives differently, and themselves differently, and their place in society differently.

Most Silicon Valley startups—more than 90 percent of them—died. But *what if* Instagram didn't? If the founders were very lucky, if they outnavigated all the competition, supported the new users, and got as big as Facebook one day, they would indeed change the world. Or at least how people saw the world, the way linear perspective changed painting and architecture in the Renaissance.

Systrom wasn't as confident as he sounded. He had actually been nervous to meet Crowley, since Foursquare was the talk of the industry. Instagram's infrastructure was still struggling to support all the new users. He and Krieger weren't sleeping well. There were plenty of strong competitors. But pretending things were going more smoothly than they actually were was part of the job of being a startup CEO. Everyone needed to think you were on the right track. His posturing was perhaps analogous to the modern pressure Instagram would introduce—the pressure to post only the best photos, making life seem more perfect than it actually was.

THE CHAOS OF SUCCESS

"Instagram was so simple to use that it never felt like work. I kept telling myself that once Instagram stops being fun to use, once it feels like work, I'll stop using it. But it stayed simple."

—DAN RUBIN, @DANRUBIN, PHOTOGRAPHER/
DESIGNER ON INSTAGRAM'S FIRST
SUGGESTED USER LIST

Mike Krieger couldn't afford to go anywhere without his laptop. He brought it to bars, to restaurants, to birthday parties, to concerts. He fixed Instagram in the back of movie theaters, in parks, and even while camping. He set up an alert on his iPhone to tell him whenever there was a spike in activity that caused the servers to fail. As Instagram started to become popular among design-obsessed Japanese users, the alert routinely woke him up in the middle of the night. The sound of the alarm—even the same sound on someone else's phone—would induce instant stress.

Not that he was complaining. It was a gift to be this busy. It meant that the app was catching on—that some of the world's new iPhone users had seen a filtered image somewhere on the internet and thought, *How do I take photos like that?* It meant that the people who downloaded the app to ride the trend had started to look at their surroundings differently.

New Instagram users found that basic things, like street signs and flower bushes and cracks in the paint of walls, all of a sudden were worth paying attention to, in the name of creating interesting posts. The filters and square shape made all the photographs on Instagram feel immediately nostalgic, like old Polaroids, transforming moments into memories, giving people the opportunity to look back on what they'd done with their day and feel like it was beautiful.

Those feelings were validated with likes, comments, and follows, as the new users built new kinds of networks on the internet. If Facebook was about friendships, and Twitter was about opinions, Instagram was about experiences—and *anyone* could be interested in anyone else's visual experiences, anywhere in the world. Krieger enjoyed capturing photos of his cat, bright lights at night, and decadent desserts; Systrom posted multiple times a day, with quick shots of friends' faces, labels of bourbon bottles, and platefuls of artisanal food.

It was a surreal moment for the founders. On the one hand, they'd built something people actually liked. On the other hand, it was possible that Instagram was just a fad. Or that the other photo-editing apps had a better strategy. Or that they would run out of money. Or that Krieger would miss his iPhone alarm.

The chaos forced Krieger and Systrom to figure out what their priorities were—whom they would hire, whom they would trust, and how they would handle the pressures of building a service that now answered to millions of strangers—laying the foundation for a corporate culture in the process. The feel-good Instagram phenomenon was theirs to ruin.

• • • • • • • • • • • •

The earliest stressors were of their own making. The app could have launched with hardier infrastructure, or a more robust set of features, but the founders hadn't known if Instagram would be popular. Krieger reasoned that if they'd spent more time building, they might have missed their moment. He thought back to the crime data app he'd helped make, with its heavily produced graphics but nobody to appreciate them. It was better to start with something minimalist, and then let priorities reveal themselves as users ran into trouble.

After launching, besides the server meltdowns, they were overloaded with customer support problems. When people couldn't remember their passwords or needed to change their usernames, there was not yet a way to fix the problem within the app. Systrom was just responding to users' tweets and giving out his email address, employing a strategy that was unsustainable. So he reached out to Joshua Riedel, a former community manager at Nextstop, which had by then sold to Facebook. Riedel, a lanky aspiring novelist, had just signed a lease in Portland, Oregon, but loved Instagram and made plans to move back to California.

Soon after, they hired their first engineer, Shayne Sweeney. Though he was only 25, Sweeney had been coding since he was a teenager, skipping college to help build web startups and later iPhone apps on behalf of various clients. He had also worked in Dogpatch Labs, where before joining Instagram he'd helped Systrom learn the Apple operating system, showing him how to build the iPhone's camera functionality into Instagram so that people could take pictures inside the app.

Sweeney was also more experienced in building app infrastructure and could help Krieger put out the server alarm fires. It was all-consuming. Once, a music venue told Sweeney he'd need to check his laptop bag at the front before going inside. He opted not to go to the concert. He was so busy he forgot for an entire month to message the woman he was dating. When he remembered to get in touch and apologize, she had already moved on.

In November, about a month after launch, the team moved out of the distracting Dogpatch Labs and into a small windowless area of Twitter's former offices in San Francisco's South Park neighborhood, which investor Chris Sacca helped secure. They took a trip to the Apple store, double-parking in San Francisco's touristy Union Square neighborhood, to purchase their first real computer monitors, which were piled atop Sweeney in the back seat for an uncomfortable few blocks' return to the office. When they settled in, it finally felt less like a project and more like a company.

With so much to get done, the founders divided and conquered based on what they were good at. Systrom was the public-facing guy, navigating the relationships with investors and the press, and working on the look and feel of the product. Krieger was behind the scenes, learning on the fly how to solve the complex engineering problems that supported Instagram's growth. Krieger, who owned much less of Instagram than Systrom did, embraced the hierarchy. He didn't want Systrom's job, and Systrom didn't want his. That's why it worked.

· · · · · · · · · · · ·

Systrom had always been good at collecting mentors and seeking their advice, just like he was good at courting interesting people to use Instagram. But now real money was involved in some of the relationships, and, to his dismay, real politics.

By the end of 2010, there were even more photo-sharing apps on the market, including PicPlz, Burstn, and Path. PicPlz, which worked on Android and had no square requirement, also came with filters, but with no preview for what they might look like before posting the image. Path, founded by an early Facebook employee, was a mobile social network for a limited number of friends, with more than just photos. And Burstn was like Instagram, but with a website.

In December, Andreessen Horowitz, the firm that had earlier invested $250,000 into Instagram, led a $5 million investment round in PicPlz. The technology blogosphere lit up with talk of conflict of interest. Was Andreessen picking the horse it thought would win the race while still being invested in the other?

Systrom was shocked to read about it. Then he got a call from a representative of the firm, accusing him of starting the negative press cycle. When he said he hadn't talked to a reporter, they asked if he was at an out-of-state technology conference, gossiping. But Systrom was at Taqueria Cancún with Krieger in San Francisco, eating a quesadilla suiza.

He was livid. The firm was investing in one of Instagram's biggest competitors and then blaming Systrom for the negative press cycle that followed. He hung up and explained to Krieger.

Krieger, with even less of a stomach for confrontation than Systrom, agreed that it was bullshit. But who in the real world cared what Andreessen thought? The only important thing was making Instagram reach its potential, he said. Only they could make that happen.

"Other people won't always be in this for us," Systrom realized. They could trust each other, and that was basically it. Nobody else was going to have Instagram's best interests in mind.

• • • • • • • • • • • • •

Being doubted was a powerful motivator. But ultimately, the Silicon Valley elite weren't going to determine Instagram's future—regular people were. Investor Steve Anderson reminded Systrom and Krieger of their strongest asset. "Anybody can build Instagram the app," he said, "but not everybody can build Instagram the community." Those artists, designers, and photographers were turning into evangelists for the product, and Instagram needed to keep them as excited as possible for as long as possible.

Twitter users for years had been independently hosting #tweetups to meet people in person that they followed online. Instagram was inspired to do the same, except they would organize the events themselves. Led by community manager Riedel, they hosted what they called an InstaMeet at Bloodhound, a masculine cocktail bar with a pool table and an antler chandelier. They publicly invited local Instagram users to meet up with the team and talk about what was working and what wasn't.

They weren't sure if anyone would come, figuring that either way, they would end up at a bar with a decent mixologist. But a crowd of about thirty people trickled in, taking up the whole space, some who knew the founders from being invited to use Instagram, others who were strangers. Members of the local press, including *TechCrunch* reporter M.G. Siegler, were there. So was Cole Rise, who had made the filters, and more recently designed a new logo for Instagram—a brown-and-tan camera with a rainbow stripe. Scott Hansen, the musician better known as Tycho, tagged along on a friend's invitation.

"Hey, man, are you Scott Hansen?" Rise asked the music artist.

"Oh, you're Colorize!" Hansen said, mispronouncing @colerise. Rise, who had gained plenty of followers as one of the first Instagrammers, was suddenly recognizable to strangers. Having an audience would eventually alter the course of his life, but for now, Rise was excited for his new friends while secretly mourning the loss of his art, as his unique way of altering photos was now available to the masses. The Hudson filter was based on the texture of the chalkboard in his kitchen, and now elements of his kitchen chalkboard were being shared the world over.

Systrom was giving him public credit, at least. The CEO had to change the name of the Spectra filter because Polaroid owned the brand name. He renamed it Rise. Cole was touched when he found out from a *TechCrunch* post. Years later, he'd launch his own filter app.

• • • • • • • • • • • •

With the meetups, Riedel wasn't just cultivating feedback—he was building a culture around the product. He thought Instagram would be stronger if people cared personally about the time they spent there and discovered other interesting people to follow beyond their friend groups. At the InstaMeets, they could talk about their amateur techniques for capturing the world's beauty. They could bask in modern creativity. There was a millennial optimism to it all. The generation that had entered the workforce during the Great Recession seemed to be saying, with every Instagram post, that they valued being interesting more than they valued the nine-to-five.

But already, there were signs that Instagram's trajectory was veering away from hipster artisans and more toward the mainstream, and corporations that weren't pretending to be anything else. By January, brands like Pepsi and Starbucks had made accounts, as had media organizations from *Playboy* to National Public Radio and CNN. Brand participation was something for any startup to celebrate, as it was the first step to a business model. But Systrom made sure to point out it was happening naturally. "We're not interested in paying anyone to use the product," he told *TechCrunch*.

The first big celebrity to sign up was the rapper Snoop Dogg. He posted a filtered Instagram picture—of himself wearing a suit and holding a can of Colt 45—and simultaneously sent it to his 2.5 million followers on Twitter. "Bossin up wit dat Blast," he wrote. Blast by Colt 45 was a new kind of fruity, caffeinated drink, clocking in at 23.5 ounces and 12 percent alcohol content.

It was the first case of an ambiguous advertisement on Instagram. Who paid Snoop to promote the drink? Or was it something he decided personally to endorse? Did it comply with advertising disclosure rules, or rules against marketing alcohol to minors?

Nobody knew, and nobody asked. A few months earlier, the FDA had warned about the danger of alcoholic drinks with caffeine in them, especially the types that came in teen-friendly flavors like grape and lemonade. But it would be years before Instagram or regulators would come up with rules about advertising disclosure on the site.

Systrom and Krieger hoped brands and celebrities would use Instagram to show behind-the-scenes content, so their posts would blend in well with Instagram's typical fare—the photos that provided windows into another person's perspective. Either way, it was nice to have a celebrity along. Stars had built communities and cultures around themselves, just like Instagram was trying to. In February, Systrom and Riedel made it to the Grammy Awards and walked the red carpet in tuxedos, Instagramming it all the while, Systrom reveling in the opportunity to get fancy.

As Siegler wrote for *TechCrunch* at the time: "Step one: obtain a ton of users. Step two: get brands to leverage your service. Step three: get celebrities to use your service and promote it. Step four: mainstream." In his estimation, Snoop put Instagram on step three, just a few months after its launch.

· · · · · · · · · · · · ·

Systrom had no trouble recovering from the Andreessen Horowitz ordeal. By early 2011, Instagram had far surpassed PicPlz in users, and his angel investors, Jack Dorsey and Adam D'Angelo, had talked him up to another respected venture capitalist.

Matt Cohler was a partner at Benchmark Capital, known for backing eBay in the 1990s, now with money in Twitter and Uber. Cohler, an early

Facebook employee before becoming an investor, thought Instagram was the first app he'd ever seen that looked like it was designed exclusively for a mobile phone, not a desktop computer. Systrom told Cohler he admired Facebook and wanted to learn more about how to build a company whose product was so ubiquitous.

Cohler agreed to invest, and joined Steve Anderson on Instagram's board. The money from the Series A investment round—$7 million, led by Benchmark—would be enough to fuel Instagram for many months, depending on how many people they hired. "We're going to grow the team to support the scale and massive growth we're seeing," Systrom told the press that February, as Instagram passed 2 million users. "We want to build a world-class engineering team."

But Instagram had four employees—Systrom, Krieger, Riedel, and Sweeney—and wouldn't hire a fifth until August. Systrom and Krieger would say they were too busy to recruit; in reality, it was hard to find people willing to leave their jobs and devote everything to Instagram. They would hear potential recruits explaining that they didn't think it was an independent company for the long term—that it was just the best way, for now, to share photos to Twitter and Facebook.

And if a candidate wasn't willing to work long hours, or didn't understand how big the vision was, Systrom would reject them, irritating investors who knew he was short-staffed. He made excuses.

"We only hire the best of the best," he told *Gizmodo* blogger Mat Honan.

That meant something different at Instagram. Systrom had been at Google, where anyone with an advanced engineering or science degree from an Ivy League school was a shoo-in, giving the place its academic feel for always running tests and optimizing. He'd also seen early Twitter, which attracted anarchists and misfits, giving the place its free speech and anti-establishment ethos. Instagram's top candidates were people with interests beyond technology, whether it was art, music, or surfing. Krieger loved talking with Riedel about literature, for example.

The painfully small team developed an in-the-trenches camaraderie. Every day, whoever purchased lunch would tend to pick it up for everyone. There wasn't much need for email. They were all in the same room, listening to Krieger's favorite indie tunes through a small speaker. They snacked on bulk orders of crumbly Nature Valley granola bars and sugar-free Red Bull, retrieved from a cubby that sometimes attracted ants. Systrom's mom sent cookies. When they had the time, they all got their hair cut by the same local barber.

The company was pushing faster, sleeker versions of the app to iPhones so frequently—once every couple weeks—that Sweeney didn't have time to write a detailed description of what was new for the Apple app store. It would be too technical, anyway. He came up with a catch-all explanation, that other Silicon Valley apps would start borrowing: "bug fixes and performance improvements."

All their nights and weekends were paying off. One day, Instagram passed Facebook in app store popularity. The milestone called for a bonus of some kind, so Systrom bought each of the employees a bottle of Black Maple Hill bourbon that cost more than $100. Sweeney, from rural Paradise, California, played a joke on Systrom's elite East Coast taste by texting him a picture pretending to pour it into his Mountain Dew.

Around the same time, at a cocktail party at the new investor Cohler's home, Systrom saw Mark Zuckerberg for the first time in years. It turned out that Instagram was on Facebook's radar too. The CEO congratulated him on Instagram's success.

• • • • • • • • • • • •

By the summer of 2011, Twitter had about 100 million monthly users, and Facebook had more than 800 million. Instagram was a much smaller player—with 6 million sign-ups—but had reached that milestone about twice as fast by building off the existing networks.

Nowhere was the effect more apparent than with celebrities. Justin Bieber had more than 11 million followers on Twitter. So when the 17-year-old pop star joined Instagram and tweeted out his first filtered photo, a high-contrast take on traffic in Los Angeles, Krieger's alarm sounded. The servers were stressed as Bieber gained 50 followers a minute.

"Justin Bieber Joins Instagram, World Explodes," *Time* magazine reported. Almost every time the singer posted, throngs of tween girls would overload the servers again, often taking them down.

Scooter Braun, Bieber's manager, had seen this movie before. All the celebrities, Bieber the most internet-famous among them, had been posting their content on social media sites and getting nothing in return. Braun had discovered Bieber in 2006 when he was barely a teenager, singing on YouTube, but he hadn't been as famous during the early rise of Facebook and Twitter. Maybe, Braun thought, he could get something out of Instagram.

The star music industry negotiator called Systrom while the latter was in a station wagon full of friends, passing by Davis, California, on the way to Lake Tahoe. "Kevin, I've got Justin on the line," he said. The two of them made a pitch: let Bieber invest, or pay him for his content. Or else he'll stop using Instagram.

Systrom had already decided that Instagram wouldn't be paying anyone for their content, since he wanted everyone to be spending time on Instagram because it was fun and useful, not for commercial reasons. He said no to paying Bieber or taking his investment.

Bieber followed through on Braun's threat. But his on-and-off girlfriend, the Disney actress and singer Selena Gomez, loved to use Instagram, and their relationship was all the gossip blogs wanted to write about. Soon Bieber was back on the app, continuing to overload Instagram's infrastructure, to the point where the company had to devote half a server just to his account's activity.

Bieber's following was enough to change the nature of the Instagram community. "All of the sudden, Instagram was emoji heaven," Rise later recalled. As younger users joined, they invented a new etiquette on Instagram, which involved trading likes for likes and follows for follows. "Instagram's community of earnest people telling interesting stories in tiny moments really evolved to be super pop culture."

• • • • • • • • • • • •

As more people joined Instagram, because of Bieber or otherwise, Riedel was hosting more InstaMeets to bring them together in real life. At one summer event in San Francisco, Rise introduced Instagram employees to one of their biggest fans: Jessica Zollman.

Zollman worked at Formspring, an anonymous question-and-answer site popular among teens. The site had turned into a cesspool of bullying, as anonymous products usually did. Teens asked their schoolmates what they *really* thought, and whoever posted was told, often enough, that they were actually nasty and ugly and didn't deserve to exist. Zollman was the one who handled the communication with the police or FBI when there was a threat of violence or suicide.

Instagram was her escape from all this. At work, they called her the "Instagram queen," making fun of her obsession. Her more artistic friends ridiculed her too, for calling herself a photographer even though she was using her phone to take photos. But she couldn't help but love Instagram, which seemed like a happier, more creative place on the internet, in a way that felt revolutionary. She was pulling together a conference focused on mobile photography, called "1197," after June 11, 1997, the day the first camera-phone photo was shared.

This was the kind of person who fit Systrom's bar for enthusiasm. Riedel, after meeting her, emailed to see if she was interested in joining the team as a community evangelist, getting other people excited about the product.

"If I change my font to 120 point size, in hot pink, with a 'hell yes!' would that be too much?" Zollman replied.

"That's exactly the right amount," Riedel wrote back. She became employee number five.

· · · · · · · · · · · ·

Systrom and Krieger were getting better at understanding their limits—or maybe they were just afraid to ruin what they had. They would not pay celebrities or brands, they would not overcomplicate their product, they would not be pulled into investor drama. They would play nice with the tech giants, they would foster community through InstaMeets, and they would try to make Instagram live up to Zollman's ideals of a friendly place on the internet.

The problem was, while Instagram could inspire users with its community efforts, it could not control them. Instagram, like Twitter, didn't require people to give their real names. Some people were less interested in 'gramming sunsets or lattes and more interested in harassing others in comments or posting content that Systrom and Krieger found objectionable.

When they'd see someone they thought was behaving badly, they'd go into the system in the "admin" page for the account, and click to block them from logging into the app, with no warning. They called the process "pruning the trolls," as if Instagram were a beautiful plant with some yellowing leaves.

Besides the bullies in Instagram comments, there were others posting graphic photos of their suicide attempts, or passing around images of child nudity or animal abuse, or posting #thinspiration content—the kind that glamorized anorexia and bulimia. Systrom and Krieger didn't want any of this to be on Instagram and knew, as the site got bigger, that they wouldn't be able to comb through everything to delete the worst stuff manually. After just nine months, the app already hosted 150 million photos, with users posting 15 photos per second. So they brainstormed a way to automatically detect the worst content and prevent it from going up, to preserve Instagram's fledgling brand.

"Don't do that!" Zollman said. "If we start proactively reviewing content, we are legally liable for all of it. If anyone found out, we'd have to personally review every piece of content before it goes up, which is impossible."

She was right. According to Section 230 of the Communications Decency Act, nobody who provided an "interactive computer service" was considered the "publisher or speaker" of the information, legally speaking, unless they exerted editorial control before that content was posted. The 1996 law was Congress's attempt to regulate pornographic material on the Internet, but was also crucial to protecting internet companies from legal liability for things like defamation. The law was the main reason services like Facebook, YouTube, and Amazon could grow very large, since they didn't have to review every hour of video that might be violent, every product review that might be disparaging, or every post that might be untrue.

Zollman knew this because at Formspring, she'd gone with her boss to a meeting with Del Harvey, the person in charge of dealing with these same legal issues at Twitter. "Del Harvey" was a professional pseudonym to protect the employee from the throngs of angry internet users she made rules for. The Section 230 law was the one thing that had stuck with Zollman from the meeting.

Still, Zollman didn't want Instagram to ignore these posts. She knew from Formspring how a dark culture could grow if untended, and how Instagram had become an escape for her. The number of Instagram users was still small enough that Riedel and Zollman could personally click through all of the damaging content to decide what to do, finishing the job in shifts. But eventually they would be overwhelmed by all the suicide attempts—and worse, she explained. After Bieber's arrival, there were young, impressionable users too.

Zollman, whose username was @jayzombie because of her fascination with the macabre, could handle seeing open wounds. But she had a big heart and refused to feel helpless. She created a basic email she could auto-send to each person she saw posting suicidal content, giving them a link to mental health hotlines in all the countries Instagram operated. She reported violent threats and other items to the police when she saw them. She established herself as the point of contact for the police and FBI, as she had at Formspring.

Getting involved didn't always produce a happy outcome. Once, she reported the suicide threats of a young Scottish girl to the police. When authorities followed up and wanted more information on the user, she couldn't help. Instagram didn't track location and couldn't give out Apple IDs, per Apple's developer terms. In another instance, as she was reporting child pornography to the National Center for Missing and Exploited Children, she found out that it was actually illegal for the company to keep the images on its server, or for her to simply email them. The folks at the center walked Krieger through what Instagram needed: a separate server that auto-destroyed content after a period of time, to enable Instagram to safely report it to authorities. Krieger set one up.

Bigger technology companies had the resources to separate their community growth work from their content cleanup work—and usually didn't devote so much attention to such things so early in their life cycle, because the law said they didn't have to. But understanding the ugly potential of the platform early helped Zollman and Riedel think not just about how to address the problems, but also about how important it was to actively promote the kind of content they wanted to see.

At Twitter and Facebook, executives reasoned that it was legally safer to be as uninvolved in content policing as possible. If there were problems, users could report or resolve them themselves, and it wasn't the company's job to tell them how to interact with the product. Riedel and Zollman saw it differently.

Because Instagram didn't have an algorithm or any way to re-share photos, there was no natural way for content to go viral. So Instagram employees had the opportunity to decide for themselves what kind of user behavior to reward, handpicking interesting profiles to highlight on their company blog. They also leaned on their users for help improving the product, asking via Instagram if anyone could volunteer to help translate the app into other languages, or if anyone could organize InstaMeets of their own around the world. They published tips for making higher-quality posts, with ideas for interesting angles and novelty perspectives, like shooting underwater.

The strategy ended up creating more and more superfans, who augmented the work Zollman and Riedel were doing, for free. Unofficial Instagram ambassadors in various countries, inspired by the company's prompts, would post about their plans to go on photo walks in picturesque spots, and then strangers would join them, exploring regions near their homes that they'd never had a reason to visit before.

The founders highlighted users like Liz Eswein, a student at New York University who had to take time off school when sick with Lyme disease her junior year. After reading a *New York Times* story about Instagram, Eswein joined early enough to grab the @newyorkcity handle, entertaining herself during recovery by taking photos of dramatic skylines, pickup basketball games, fish markets in Chinatown, and street performers she observed while walking around. To help the app grow, she posted about Instagram meetups and scavenger hunts to bring local users together at public parks or bars and simply observe their city through their phones. And in turn, Instagram's promotion of her account was helping her gain 10,000 followers a week.

· · • · · · · · · · · · •

Adding a re-share button would give Instagram less power to demonstrate model behavior; everyone would just be focused on going viral. Still, their users seemed to be asking for one. Twitter had just added a retweet button, to account for the fact that users were copying and pasting each other's tweets naturally. Having an automatic way to share posts would be great for growth. Besides rewarding users with virality, the option to share others' content could lessen the pressure on people who didn't feel they had anything interesting to photograph.

Krieger did build a re-share button but never released it to the public. The founders thought it would violate the expectations you had when you followed someone. You followed them because you wanted to see what *they* saw and experienced and created. Not someone else.

The founders would constantly need to defend this thesis, now that social networking was synonymous with virality. And it wasn't just Silicon Valley types who were asking.

By September 2011, Instagram had 10 million users. Hollywood luminaries were still trying to invest, traveling to the company's tiny section of its South Park office. The actor and singer Jared Leto made a desperate pitch: "You mean if I left you a bag of money at your doorstep right here, you wouldn't take it?"

Ashton Kutcher, the actor from *That '70s Show* and comedic movies like *Dude, Where's My Car?*, in 2009 beat CNN to have the first Twitter account with 1 million followers. Like Bieber, he realized that he was creating a lot of value for that company for nothing in return. Kutcher was nothing like the dopey characters he played. He absorbed everything he could about the technology industry, resolving to use his trend-spotting skills in a more lucrative way. He worked with Guy Oseary, Madonna's manager, to sort through all the opportunities, and ended up giving money to dozens of companies—not just in social media—including Uber, Airbnb, Spotify, and Instagram competitor Path. "Whenever there was a new type of experience for consumers, there would be like three companies doing the same thing," he remembers. There were several versions of Instagram, Pinterest, and Uber. "Who would get traction first? And then the network effect would take all."

To know if Instagram was a fad or a lasting network, Kutcher and Oseary looked at data that showed users were spending more and more time there, building a habit. "It's a competition for attention," Kutcher explained. "Everybody learned that from Facebook and Twitter."

Oseary and Kutcher struggled to get a meeting. But eventually, they made it to the South Park office, with its brown carpet and 1980s glass-block windows that barely let any light inside. There, they found a busy team immersed in their screens, trying to keep the app from crashing, too busy to talk.

Systrom stepped aside and explained that he wasn't looking for new investors, but was willing to explain Instagram's market opportunity. Filters made sharing photos easier, lowering pressure on users. A filter on Instagram was like if Twitter had a button to make you more clever. "If I can help people make those photos beautiful, it makes them more shareable, and by making them more shareable, this thing wins," he said.

"Then you need a re-gram feature," Kutcher said.

Systrom tried to explain. "It has to be a simple, clean stream. You'll still be able to find content, but it has to always be directly attributable to its creator," he said, using an argument he thought would appeal to a person paid for his talents.

Kutcher was put off by Systrom's lack of flexibility in the face of a good idea. But he was still intrigued enough to invite him, with their mutual friend Joshua Kushner, on a ski trip in Utah with other technology founders. A half dozen men stayed there overnight in a large cabin in the snow.

In the middle of the night, Systrom burst into Kutcher's room. They had to get outside—*immediately*. Kutcher's room was already filling with smoke. The wall by the fireplace was up in flames.

Systrom ran from room to room at around four in the morning, until all the guests made it out safe. They all stood outside in the cold, in their underwear and clutching their laptops and phones dearly, waiting for the fire department to come.

Okay, Kutcher thought, *Kevin's a good leader.* They became friends, and Kutcher would later help Instagram build more credibility in the entertainment industry.

• • • • • • • • • • • • •

All of the things Instagram was doing well—getting celebrity attention, building communities around interests, becoming a natural accompaniment to life with a mobile phone—were also priorities for Twitter. The companies' destinies seemed so intertwined that when some celebrities visited Instagram, they asked if they could meet Twitter next, not realizing they were separate services.

At the end of 2011, Twitter deals employee Jessica Verrilli made the case that they shouldn't be. Instagram had built a network off Twitter's with some of the same basic structure and investors. Verrilli, who had worked with Krieger back at Stanford's Mayfield program, urged Dorsey to contact the founders again about the idea of an acquisition. Dorsey said Systrom sounded enthusiastic, as long as they could come up with an appealing number.

Ev Williams's objections wouldn't be a problem this time. A year earlier, Dick Costolo, Twitter's chief operating officer, had been promoted to CEO, replacing Williams, who'd stepped into a product role. By March 2011, Dorsey had convinced Twitter's board that he was the rightful visionary to lead Twitter's future as executive chairman. The board pushed Williams from his perch with the help of Benchmark investor and board member Peter Fenton, just as they'd pushed Dorsey out in 2008. Dorsey, while still running Square, came in as Twitter's executive chairman, working with Costolo to lead the product direction.

When there are competing visions at the top of a company, the executives often fight for recognition of their own relevance and impact, getting in the way of doing what's best for their consumers. This is what

Twitter employees observed in their management. Costolo wanted to assert himself as CEO, but Dorsey was the founder, and so they jostled for the spotlight.

The number the deals team had in mind for Instagram was around $80 million. Costolo said that $80 million was far too expensive for such a young company that was just a better way to share photos within Twitter. It was unlikely to compete with Twitter on being a resource for news and updates from public figures. Others in the room guessed there was another element to his thinking: if the deal came to pass, Dorsey would get credit for it. The discussion was tabled.

But Instagram kept getting more valuable, finding its footing and a path to the mainstream. Despite Costolo's doubts, celebrities continued to sign onto Instagram, including Kim Kardashian, Taylor Swift, and Rihanna. In January 2012, Instagram added one of Twitter's most valuable users: President Barack Obama. Obama's account launched the day of the Iowa caucus for that year's presidential campaign.

Instagram said on their blog that they wanted the account "to give folks a visual sense of what happens in the everyday life of the President of the United States," and asked news reporters to join in posting behind-the-scenes moments from the campaign trail. That same month, Krieger was invited to be Michelle Obama's guest at the State of the Union, to explain that he wouldn't have been able to help start Instagram without an immigration visa.

Meanwhile, the Instagram team had finally been growing. Amy Cole, who in a past life consulted with race car drivers on their aerodynamics for Chrysler, had just graduated from Stanford business school. A friend told her he could introduce her to the Instagram team after hearing her rave about the app on a wine-tasting trip in Napa. She became their first head of business in October 2011, though there was no real business to conduct yet. She helped the company find a long-term lease in a bigger building across the street, with real windows. Gregor Hochmuth, the friend Systrom had initially asked to be his cofounder, came on as an engineer in December to make more filters, which seemed like the killer feature at the time.

The company also formalized some of its editorial initiatives. Instagram had an opinion about what model content on the site looked like—the kind that gave a window into an interesting life. Bailey Richardson, a member of the community team starting in February 2012, curated a list of "suggested users" for people to follow, so they didn't by default think Instagram was about celebrities. It included photographers, artisans, chefs, and athletes around the world. The list especially featured those who had been diligent about appearing at or organizing InstaMeets, like Eswein of the @newyorkcity account. Richardson also found and promoted characters like @darcytheflyinghedgehog, an account run by a young Japanese man who liked to dress up his tiny hedgehog, and @gdax, a Tibetan monk calligrapher.

The technical tools to manage accounts were still scrappy—or nonexistent. One day that winter Scooter Braun sounded an alarm to Instagram: Bieber was locked out of his app. But Instagram didn't have a reliable system for resetting passwords. They told Braun they could do it over the phone, but Bieber would need to verify his identity. "All right," Braun said, "Justin's going to call you guys."

Richardson picked up. "Hey, this is Justin," he said. There were no security questions prepared, so that declaration would have to be enough proof of identity. She reset his password over the phone.

• • • • • • • • • • • • •

By the beginning of 2012, a longtime Twitter employee named Elad Gil took over Twitter's corporate strategy and M&A. He resurfaced the idea of an Instagram acquisition. Important people were joining the app, and things were starting to *happen* there, he explained in a presentation on his strategy for the quarter. In 2009, Twitter had started to be taken seriously as a news source because someone posted an incredible photo there of a plane perfectly landing on the Hudson River in New York. What if the next photo like that was posted on Instagram instead? What if Instagram became the default way to share photos? The scenario would be bad for Twitter unless Instagram joined, Gil argued.

Dorsey, still executive chairman, was no longer involved in the day-to-day product work. And this time, Costolo was not only receptive to the Instagram idea but willing to be very aggressive. He met with Krieger and Systrom at the bar of the Four Seasons hotel in San Francisco.

The Instagram founders were not overly enthusiastic because they felt they were just getting started. Systrom thought it was good practice to be polite and meet anyway. Twitter held the keys to so much of Instagram's growth. So did Facebook. So did Apple. If they harmed any of these relationships, they could harm the company's potential.

Costolo left the meeting thinking that if he could woo them enough, he could make it happen. Around the same time, Dorsey was trying a different angle, inviting Systrom and Krieger for a casual catch-up at Square's offices. Costolo and Dorsey were aligned: Twitter needed to bet big on this one. The number needed to look insane, but it would be worth it.

So Gil and Ali Rowghani, Twitter's chief financial officer, drew up a term sheet for the acquisition. Twitter was willing to offer between 7 and 10 percent of its stock for the deal, worth between $500 and $700 million, with some interpretive wiggle room since Twitter's shares weren't public yet. The percentage was calculated based on the idea that Instagram had between 7 and 10 percent of the 130 million users Twitter had.

In March, Systrom gave a presentation at an exclusive conference hosted in Arizona by the investment bank Allen & Company. Rowghani, Costolo, and Dorsey were there. Late one afternoon, a couple days into the conference, Rowghani and Dorsey met Systrom for drinks around a patio fire pit. Dorsey wasn't drinking at the time, but Systrom sipped whiskey.

Nobody agrees about what happened next. Twitter sources say Rowghani presented the term sheet, with a place for Systrom to sign, and Systrom handed it back, saying he didn't think he should sell. But Systrom would later deny he heard numbers or saw a slip of paper.

Term sheet or not, everyone agreed that Systrom hadn't accepted. Twitter launched an all-out offensive to wine and dine the founders until they could be convinced.

THE SURPRISE

"He chose us, not the other way around."

—DAN ROSE, FORMER VP OF
PARTNERSHIPS AT FACEBOOK,
ABOUT KEVIN SYSTROM'S DECISION

Gregor Hochmuth needed a few seconds to answer his phone because he was engrossed in the challenge of eating his dinner—an enormous burrito from San Francisco's Mission District, with a tortilla stretched tight around an unwieldy pile of ingredients. Any mishandling could cause dollops of guacamole and salsa-soaked rice to roll out.

The call was from Krieger. It was rare for Hochmuth to hear from his boss so late on a Sunday night.

"Everything okay?" Hochmuth asked.

"Hey, man," Krieger said. "You need to come into the office early tomorrow."

Hochmuth had been spending almost every waking moment at the office. The prior week, on April 2, he had spent the night to help prepare for the launch of Instagram's Android app.

"I usually get there at like eight," the engineer said, feeling defensive.

"Eight. Eight is good," Krieger said. There was some news to discuss, he explained, and then hung up.

Hochmuth spent his next few messy bites wondering what it could possibly be.

• • • • • • • • • • • • •

Later that night, Tim Van Damme was driving down a mountain, feeling grateful. He was finally in California, ahead of his first full week in San Francisco with the Instagram team. They had hired him over the winter when he was pretty desperate. His Austin, Texas–based employer, the location check-in app Gowalla, had been acquired by Facebook in December, but not all employees had been brought in with the deal. Without Instagram, he would not have had health insurance in time for his wife to give birth to their first daughter.

Van Damme got lucky because Systrom happened to read a Twitter direct message he sent, praising the Instagram product and wondering if they needed any help. It turned out that they did need help—desperately. Systrom and Krieger had been so busy, they hadn't had time to start looking for a designer. He did a couple job interviews with the founders. Krieger had to interrupt their conversation at one point to reset the servers, because the teenage heartthrob Justin Bieber had posted again, causing them to crash. That was a fun problem to have, Van Damme thought.

The designer became Instagram's ninth employee when his daughter was just a couple days old. Most startups fail, he thought, but at least for now he had a job, redesigning buttons and logos on an app whose creators cared about style. And he got to recommend a friend too: Instagram hired Philip McAllister, another Gowalla employee who hadn't gotten a Facebook offer, to engineer the Android version of the app.

Van Damme worked from his tiny kitchen table in Austin until his young family could make the move. Three months later, when they got to California, they celebrated the transition with a weekend in Lake Tahoe, hoping to capture some late-season snow that Easter weekend.

As he was making the three-hour drive back to their new home, Van Damme's phone rang. It was his CEO.

"Can you be at the office tomorrow at eight a.m.?" Systrom asked.

"Okay," Van Damme said.

"Thank you," Systrom said. "Have a good night." That was it.

Van Damme took his eyes off the road for a second to shoot a look of panic at his wife.

"I'm going to be fired," he explained. He knew it in his bones. "Nobody does a meeting at eight a.m. in Silicon Valley."

* * * * * * * * * * * *

When Van Damme and Hochmuth got to the office the next morning, it was clear that everyone else had gotten the same message. The employees whispered their theories to one another. Maybe there had been a major hack. Maybe something had gone wrong with the recent venture capital fundraising, and Instagram was actually out of money and would have to shut down.

In the front room of the new South Park office, they arranged a half circle of chairs facing the door. Josh Riedel dialed up Dan Toffey, their one employee in Washington, DC, and slid his iPhone across the navy-blue carpet toward Systrom's shoes, so he could also hear whatever the founders had to say.

"So over the weekend, we had some conversations about a potential acquisition," Systrom said.

That wasn't that crazy, the employees thought. Last week's Android app launch had been pretty successful, downloaded 1 million times in the first 12 hours.

"I talked to Mark Zuckerberg," he continued.

Still normal.

"We said yes to Facebook. We're getting bought—for $1 billion."

Not normal. Not believable.

Employees let out gasps and guttural sounds. Some of them laughed, unsure how to control their surprise, while others failed to hold back tears. Jessica Zollman grabbed Hochmuth's thigh. Amy Cole grabbed the hands of the people sitting next to her. Tim Van Damme and Philip McAllister made eye contact. Anyone but Facebook, they thought. It'll be the Gowalla situation all over again.

But it was $1 billion. *One billion*—a magic number, unheard of for mobile app acquisitions. Google had bought YouTube for $1.6 billion, yes—but that had been six years ago, before the U.S. financial crisis. Facebook didn't do acquisitions like this. Everything they bought, they stripped for parts, keeping the founders and technology but killing the product. Was Instagram going to be killed? Did they need new jobs? Or was it possible they had all come into some serious wealth? Shayne Sweeney was nervously ripping off the label of his empty Perrier water bottle, stuffing the scraps inside.

Systrom explained what would happen next, logistically. They were about to go to Facebook headquarters and meet Facebook management. A shuttle would come pick everyone up in the afternoon. But his words were barely registering. His voice had turned into background noise for the thought process going on in everyone's head, as undecipherable as the speech of the teacher in Charlie Brown cartoons.

Everyone was jolted back into reality, though, when Systrom explained that the news would become public in thirty minutes.

"Call your families," he said. "Do whatever you need to do before then."

Riedel picked up his iPhone, lying at Systrom's feet. It turned out the phone hadn't been on speaker mode, so Toffey hadn't been able to hear anything. That was the first person he needed to explain this to.

The rest of them walked back to their desks—the Ikea desks for the new office that Zollman had constructed just a month earlier. They passed some still-unopened champagne from Sequoia Capital, meant for celebrating the $50 million funding round that had closed just the previous week. That news had felt like a milestone, and this felt like a ton of bricks.

"I didn't think this was how it was going to end," Sweeney said to Hochmuth.

The questions from family were the obvious ones, about money, but there were no answers for them. Systrom hadn't addressed the topic. A few minutes later, Van Damme, still processing, needed a cigarette. He started walking toward the door.

"Don't go out there!" a coworker yelled at him from behind. It was about 9:10, which meant the news had been public for ten minutes. Robert Scoble, the technology blogger, was pulling up in front of the office in his white Prius. *Good call,* Van Damme thought, as he closed the door quickly. Getting interrogated by a Twitter personality was pretty much the opposite of what he needed at that moment.

White news vans followed, as well as photographers. Unable to escape their office, the employees looked at the Internet.

One billion dollars, Reuters said, "was stunning for an apps-maker without any significant revenue." Zuckerberg was "paying a steep price for a startup that has lots of buzz but no business model," CNN echoed, comparing the deal to Yahoo!'s $35 million acquisition of Flickr seven years earlier.

Some of Instagram's 30 million users were tweeting a different concern: that Facebook would dissolve Instagram, or incorporate it into the news feed, or simply put too much of its own stamp on the product, crowding it with features that ruined the simplicity. Meanwhile, Facebook would get control over all of their Instagram photo data—which didn't sound good. Facebook was already notorious for its tendency to change users' privacy terms, collect and share data with app developers in ways users didn't understand, and even tag people in photos automatically using software that recognized their faces.

The public statements from Systrom and Zuckerberg attempted to reassure them.

"It's important to be clear that Instagram's product is not going away," Systrom said on the Instagram blog.

"We're committed to building and growing Instagram independently," Zuckerberg's Facebook post said. "It's the first time we've ever acquired a product and company with so many users. We don't plan on doing many more of these, if any at all."

Whatever happened next was in uncharted territory—for Instagram and for Facebook.

• • • • **•** • • • • • • •

One month earlier, Twitter's romancing had been aggressive but unsuccessful. The founders were wined and dined over sushi with Benchmark partner Peter Fenton, as well as breakfast at the St. Regis hotel. CEO Dick Costolo explained his vision: that Systrom would get to run Instagram still, but could also be head of Twitter's product, and help Twitter become a more visual destination.

Systrom's lack of enthusiasm was palpable. He arrived an hour late to the breakfast, blaming the rain, leaving Krieger to entertain Costolo and CFO Ali Rowghani. The CFO, who finished his egg-white omelette before Systrom arrived, found the late arrival smug and insincere, acting like he was Hollywood talent, wasting their time.

Twitter wanted to finish the deal before the South by Southwest technology conference, which was running the week of March 9. That was the conference where Foursquare had gained major buzz in 2009 and Twitter in 2007. But Systrom stalled. At the conference, a small team of Instagram employees passed out stickers with the Instagram logo and T-shirts with dinosaurs on them. One night, at a bar, people came up to Systrom, recognizing him as the founder of Instagram and telling him how much they appreciated the product.

Back in San Francisco, Systrom told Dorsey about the gratifying experience and explained that he couldn't sell now. He wanted to make Instagram so big and important, it would be too expensive to be acquired by anybody. Dorsey said he understood. He introduced Systrom to Roelof Botha, a partner at Sequoia Capital, who started negotiating to put the venture firm's money into Instagram.

Systrom would tell his friends that Twitter never made a serious offer. In reality, they never offered him anything he wanted to take seriously. Only Zuckerberg understood what would appeal to Systrom: independence.

• • • • **•** • • • • • • •

The road to the Facebook deal started the first week of April. Sequoia was going to back a $50 million venture round at a $500 million valuation, close to the Twitter offer price, and all Systrom had to do was sign the papers. But first, Zuckerberg called.

"I've thought about it and I want to buy your company," Zuckerberg said, getting straight to the point. He wanted to meet as soon as possible. "I'll give you double whatever you're raising your round at."

Systrom wasn't sure what to do. He panicked and called his board.

Matt Cohler from Benchmark told him that whatever happened with Zuckerberg, he needed to sign the papers for the round of venture capital, or his reputation in Silicon Valley would never recover. Steve Anderson, the other board member, was stuck in a meeting in Seattle. Systrom called again and again, until he picked up.

"Mark Zuckerberg wants to meet today," Systrom said. "What do you think?"

"Look," Anderson reasoned, "you just raised money. A lot of money. And if the current king of the internet wants to meet you . . . sure, why not? There's little reason not to take a meeting like that."

Anderson had been telling Systrom that he was just as much of a visionary leader as Zuckerberg, maybe even smarter. Over time, as Instagram grew, that would become clear to everyone else, Anderson thought. He didn't think Instagram should sell—at least not yet. But for now, he might as well go kiss the ring.

Systrom signed to finalize the Sequoia round, and then called Zuckerberg back.

· · · · · · · · · · · · ·

As Facebook was publicly gearing up for its initial public offering, which would be one of the biggest in internet history only a few weeks later, Zuckerberg was forced to think about the long-term realities of his business. Facebook had made one of the most ubiquitous internet services, but their users were moving over to mobile devices fast. Facebook had an app, but, unlike Google and Apple, it didn't make phones. That meant that unless Facebook rushed into the expensive, complicated hardware business, Zuckerberg would forever be building his company inside territory ultimately owned by other companies.

Which left only two ways to win. One, his engineers could make Facebook so entertaining and useful that it took up more and more of people's time on their phones. And two, he could buy, copy, or kill competitive apps, making sure there were fewer opportunities for other companies to encroach on anyone's Facebook habit.

When he heard about Instagram's $500 million valuation fundraise, he realized that this tiny, buzzy competitor, flush with new cash, could quickly become a greater threat. The only answer was to buy it.

Zuckerberg had already tried this before—unsuccessfully, back in 2008, when Twitter CEO Ev Williams had indicated he would accept an offer worth about $500 million. But then Williams got cold feet, and now Twitter was a major competitor. Zuckerberg was upset about the outcome, but had done the same thing himself once. In 2006, when Facebook was about Instagram's age, Yahoo! had offered him $1 billion. He went against the advice of his board and said no, confident that he could build Facebook to be bigger on his own. Zuckerberg derived much of his confidence from that pivotal moment of defiance. It affirmed that a founder's instincts—his own instincts—should be trusted above all else.

Armed with those experiences, Zuckerberg thought he knew how to talk to Systrom, founder to founder. Systrom didn't want to run a Facebook product, just like he didn't want to run a Twitter product. He wanted to keep his company, and to keep being the Instagram visionary, just with none of the risks of independence. Facebook's network was already helping Instagram grow—and if Instagram was part of Facebook, they'd have unimaginable resources to keep growing, faster.

This argument seemed to appeal to Systrom. But it would take some serious negotiating: that Thursday night, at Zuckerberg's new home in the tree-lined Crescent Park neighborhood of Palo Alto, Systrom started out by asking for $2 billion.

· · · · · · · · · · · · ·

Zuckerberg was whittling down the number with Systrom when he decided to loop in others. He invited Facebook COO Sheryl Sandberg and CFO David Ebersman over for a serious meeting. They told him they trusted his instincts, but first they would need to alert deals director Amin Zoufonoun, who could make everything happen.

"Mark would like to buy Instagram," Sandberg explained on their conference call, getting straight to the point.

A wonderful choice, Zoufonoun thought—it had been on his radar since he joined Facebook from Google as director of corporate development a year earlier, and he remembered Systrom from his time on the deals team.

"He's already spoken to Kevin and they've converged on a price range at a high level," she continued. He wanted to make a deal that would value Instagram at about 1 percent of Facebook.

Zoufonoun was shocked silent. Facebook's private market valuation, a month before its planned IPO, was about $100 billion. That would mean an Instagram deal worth $1 billion. Nobody had ever paid that for a mobile app before.

"You seem skeptical," Sandberg observed. "I'll call you later tonight once you've had a chance to think about it and do some analysis."

Zoufonoun thought about it, but still couldn't make the math work in his head. Usually there are similar deals to compare to, or a public company's value to match against. When Sandberg called back, Zoufonoun asked for clarification.

"The price is really huge," he said. "I'd love to understand where Zuckerberg is coming from on this—how did they arrive at that number?"

Sandberg conferenced Zuckerberg into the call, who suggested he and Zoufonoun meet in person the next morning.

That night, Zoufonoun couldn't sleep. He had just moved with his wife and two little kids into an old house in Los Altos, the next town over from Palo Alto. He'd never done a deal this big, and his nerves were getting to him. He passed the hours until his meeting with Zuckerberg with his phone in hand, scrolling through Instagram, trying to predict its future.

In the darkness, he realized this wasn't just an app for people to post pictures of their meals, but a potentially viable business. The hashtag system for organizing posts by topic made it almost like Twitter, but visual, so you could see what was going on with a particular event just by clicking. He also saw that even though the app had a mere 25 million registered users, compared to Facebook's hundreds of millions, businesses were already using Instagram to post photos of their products, and their followers were actually interacting and commenting.

Instagram didn't make money yet, but Zoufonoun surmised that because the Instagram product provided its users with the ability to endlessly scroll through posts, just like the Facebook news feed, they could eventually develop the same kind of advertising capabilities. They could use Facebook's infrastructure to grow faster, the way YouTube did at Google.

The next morning, in a conference room at Facebook headquarters, Zuckerberg and Zoufonoun appeared as scheduled.

"Hey, what's up?" Zuckerberg asked him. "I understand you have concerns."

"Actually, after the last twelve hours, I think your instinct is spot-on," Zoufonoun concluded. "We absolutely should buy this company."

"Okay, so what's next?" Zuckerberg said, unsurprised at being right. "We should probably do this quickly. How quickly do you think we can get this done?"

Zoufonoun got up, went to the whiteboard in the conference room, and started writing out the steps: convening the lawyers, figuring out the details of the cash and stock in the payment, and determining how much risk Facebook was willing to take on by shortening the timeline for due diligence. Often companies spend weeks or months evaluating a prospect, the way a home buyer checks for termites or faulty plumbing before closing a deal. But if Facebook hustled, they could get this done in a single weekend, without any outside bankers.

Zuckerberg wanted to hustle. He was one of Silicon Valley's greatest chess players, thinking several moves ahead. If Facebook took too long to negotiate, Systrom would start calling his friends and mentors.

Zuckerberg knew, from his former employee Cohler on Instagram's board, that Systrom was close with Twitter's Dorsey. Zuckerberg didn't have the friendship advantage. But the faster he made the deal, the less likely Systrom was to call someone who would give advice unfavorable to Facebook—or a counteroffer.

Zoufonoun canceled his spring break vacation with his family.

• • • • • • • • • • • •

As the lawyers were hammering out details at Facebook headquarters, Systrom went with Krieger so he could meet Zuckerberg for the first time. Afterward, the two of them sat for about an hour at the Palo Alto Caltrain station, talking about the gravity of the decision.

Without Facebook, Instagram would have to very quickly grow its team and its infrastructure if it were to have any hope of delivering a return for the new investors—and meanwhile, there was a chance it wouldn't work, or that Facebook would perfect its own version of Instagram. Krieger deeply respected Facebook's engineering team. If they joined Facebook, they would have the resources to reach a lot more potential users, with support, so that there would be fewer service outages.

• • • • • • • • • • • •

The discussions continued at Zuckerberg's sparsely furnished $7 million home that Saturday. Zuckerberg, Zoufonoun, and Systrom were sitting in the covered backyard, along with Zuckerberg's mop-like Hungarian sheepdog, Beast. Systrom occasionally wandered to the yard or to his car, taking private calls with his board.

Krieger stayed in San Francisco, but spent the weekend handling Facebook's appraisals of Instagram's technical infrastructure. He answered questions over the phone about how Instagram's systems were architected, and what kind of software and services the company used. Facebook never asked to look at the code. *We could be running this company on Legos and they wouldn't know,* Krieger thought.

In Palo Alto, there was a disagreement about the cash versus stock portion of the deal. Cash in hand is tough to pass up compared to the riskier move of potential future gains. Zuckerberg was working to convince Systrom that with stock, the price of the deal would be worth much more in the future. One percent of Facebook was only worth $1 billion if you thought Facebook's growth was flatlining. But Facebook was planning on growing, making that stock worth closer to Systrom's original number, and beyond.

But Zuckerberg even admitted to Systrom that he was surprised at Facebook's $100 billion private market valuation. While he thought Facebook would continue to grow, and that it was fair to base Instagram's price off Facebook's, he was concerned over the acquisition price tag. If he valued Instagram so highly, with its tiny team and no revenue, he would start a bubble in Silicon Valley, raising the price of every related company he might want to buy in the future. (He was partially right. In 2013, venture capitalist Aileen Lee came up with a name for startups with billion-dollar values: "unicorns." At the time, there were 37. When she wrote an update on the rare breeds in 2015, there were 84. By 2019, there were hundreds. But if it's a bubble, it hasn't burst.)

The whole time they talked, the big white mop dog wandered around, making eye contact, rolling on his back, as if he wanted to be part of the deal.

"Are you guys hungry?" Zuckerberg asked. It was already 3 p.m. and they had only been drinking beer. "I'll fire up the barbecue."

Zuckerberg went to his freezer and pulled out a large slab of venison, or maybe boar—something with lots of bones. "I don't know what meat this is but I think I hunted it at some point," he said. The prior year, it was Zuckerberg's goal to only eat meat from animals he'd killed himself.

Zoufonoun stood next to Zuckerberg as he tended the meat, smoke billowing out of the grill. Beast looked on intently, then started growling. Suddenly the dog ran 20 feet, his corded coat flopping in the wind, lunging to attack Zoufonoun's leg. "Shit!" Zoufonoun exclaimed.

"Did he break skin?" Zuckerberg wanted to know. "Because if he did, you know, we have to record it. And they might take him away."

Thankfully Beast hadn't broken skin and Zoufonoun never did bleed, but he would later tell the story at Facebook meetings, joking that

Zuckerberg was more concerned about his dog than about his deals guy on the eve of a historic transaction.

Whatever meat Zuckerberg grilled, it wasn't satisfying. Systrom excused himself for a date with his girlfriend a couple hours later. Zoufonoun shot Zuckerberg a look: *Why is he leaving for dinner in the middle of this?* There were still quite a few contentious discussions ahead.

After dinner, Systrom drove back south to meet with Zoufonoun alone. The vibe at Zoufonoun's house was different than that at Zuckerberg's bright modern digs. The family room was a converted garage with a low ceiling, drafty windows, and parquet wood floors from the 1970s, so dark that Zoufonoun's kids called it his man cave. The men sat on large couches opposite one another, laptops open, drinking scotch, as they negotiated into the night.

Looking at Instagram's history of investor financing restored Zoufonoun's respect for Systrom. Here was a man who just a few years ago had been helping on Google's deals team, making PowerPoint presentations. He had done all this in just eighteen months.

· · · · ● · · · · · · · ●

On Sunday, Michael Schroepfer, Facebook's head of engineering, was in Zuckerberg's kitchen with Zoufonoun while Systrom paced outside in the yard, on the phone with his board.

Usually, when Facebook acquired a company, they found ways to absorb the technology, rebrand the product, and fill some gap in what their own company was capable of. If Instagram was going to be its own product, it broke Facebook's normal acquisition process, and it wasn't clear how it would work.

"How do we integrate something like this?" Schroepfer asked.

"Schrep, we are buying magic. We're paying for magic. We're not paying $1 billion for thirteen people. The worst thing we could do is to impose Facebook on them prematurely." After hours of discussion and sleepless nights, Zoufonoun was fully converted into an Instagram believer. "It's blossoming, and you just need to nurture that plant. You don't need to trim it or shape the plant at that point."

Zuckerberg agreed. He fired off an email to the Facebook board, letting them know what was happening. It was the first they were hearing of the massive deal, which was all but completed. Because Zuckerberg held the majority voting power in the company, the board's role was merely to put a rubber stamp on his decisions.

* * * * * * * * * * * * *

Systrom's board conversations faced more resistance. Anderson, in particular, was confused and opposed. Just a week ago, Systrom had been raising money to grow the company for the long term. And a month ago, he'd been rejecting Twitter.

"What's up with this change of heart?" he asked, with Systrom on the phone from his car in Zuckerberg's driveway. "If it's about the money, I know I can raise you money at whatever valuation Zuckerberg's willing to pay." Anderson thought Facebook was undervaluing its per-share price to make the deal sound less insane, and that really it was worth $1.2 billion or $1.3 billion. But taking Instagram out as a competitor could be worth $5 billion to Facebook if they just waited a little longer.

Systrom gave four reasons. First, he reiterated Zuckerberg's argument: that Facebook's stock value was likely to go up, so the value of the acquisition would grow over time. Second, he'd take a large competitor out of the picture. If Facebook took measures to copy Instagram or target the app directly, that would make it a lot more difficult to grow. Third, Instagram would benefit from Facebook's entire operations infrastructure, not just data centers but also people who already knew how to do all the things Instagram would need to learn in the future.

Fourth, and most importantly, he and Krieger would have independence.

"Zuckerberg has promised me that he will let us run Instagram like a separate company," Systrom said.

"Do you believe that?" Anderson asked skeptically. He'd seen enough buyers say whatever they needed to say to get a deal done, then renege later.

"Yes," Systrom replied. "Yes, I really do believe that."

If he was confident, Anderson wasn't going to stand in his way. At least they all believed in Facebook stock. Cohler had told them the company was run like a machine. Cohler, the former Facebook employee, had been taking these calls from a vacation in Sweden, talking to Zuckerberg and then Systrom and then Zuckerberg again, all through the night.

Back in Palo Alto, the terms were pretty much settled in time for Zuckerberg to host a small evening gathering of friends to screen that night's episode of *Game of Thrones*. Systrom didn't watch the show. He signed the contract late that evening in Zuckerberg's living room. With an oversize cursive *K* and *S* in his signature, the "Systrom" part looked like a star.

• • • • • • • • • • • •

The structure of the Instagram acquisition—a company purchased *not* to be integrated—would become an important precedent in technology M&A, especially as giant companies got even gianter, and small companies like Instagram wanted to find some alternative to competing with them or dying. In the coming years, Twitter would buy Vine and Periscope, keeping the apps separate and the founders in charge, at least for a little while. Google would buy Nest, keeping it separate. Amazon would buy Whole Foods, keeping it separate. And so many corporate development teams would court startups, promising to "do it like Instagram," only to change their minds about granting independence once everyone was in the building.

Instagram's perceived independence at Facebook would help Zuckerberg win some otherwise impossible deals with headstrong founders, especially in 2014, with the chat app WhatsApp and the virtual reality company Oculus VR.

But mostly, the Instagram deal would give Zuckerberg a tremendous competitive advantage. One Facebook executive would later reflect on the relative importance of the deal: *Imagine an alternate reality, in which Microsoft buys Apple while Apple is still small. That would have been tremendous for Microsoft. And that's what Facebook got with Instagram.*

It's an imperfect analogy. Still, the biggest challenge of such a merger is not in maintaining growth and longevity for the products, but in navigating the egos of their creators and the separate cultures of their companies. In the imagined scenario, would Microsoft get to take credit for the iPhone? How long would an eccentric creative like Apple's Steve Jobs last in a more bureaucratic corporate environment?

Zuckerberg wasn't sure how things would play out. But his motivation is outlined in a little red-orange book, handed down to new Facebook employees at every Monday morning orientation. On one of the last pages, against a navy backdrop, there are a few sentences in light blue writing that explain Zuckerberg's paranoid leadership: "If we don't create the thing that kills Facebook, someone else will. The internet is not a friendly place. Things that don't stay relevant don't even get the luxury of leaving ruins. They disappear."

The question that Systrom would be asking, six years later, was whether Zuckerberg considered Instagram part of the "we," or the "someone else."

* * * * * * * * * * * * *

The morning after Systrom signed the contract, Dorsey was on his way to work at Square, the payments company he'd cofounded. Despite his wealth, he always enjoyed taking public transportation to absorb the culture of San Francisco. That morning he noticed that he'd have the whole Route 1 Muni to himself. "A simple morning pleasure: an empty bus," he posted on Instagram, with a picture of the brown-and-tan seats of the car, unfilled, and not even the driver visible.

He'd been posting once a day, sometimes twice a day when inspiration struck, about Square and sunsets, coffee, and air travel. Despite Instagram's recent rejection of Twitter's overtures, Dorsey had only become more invested in the app's success after helping a couple investor friends get in on their latest funding round.

As he entered Square's headquarters, one of his employees asked him if he'd heard the news: Instagram had been acquired by Facebook.

Dorsey needed to figure out if it was true. He pulled out his phone to Google it and found Zuckerberg's post. Before he could process how betrayed he felt, his phone rang. It was Aviv Nevo, a close friend and introverted Israeli-American technology investor whom he'd advised to put money in Instagram's latest round, through Thrive Capital.

"I don't know what happened," Nevo said. "I just closed this Instagram thing at a $500 million valuation, and now I'm reading that it was bought for $1 billion. What does that mean for me?"

"Well, I mean, you just doubled your investment in a couple days," Dorsey said, speaking slowly, trying not to show his bewilderment. "That's one of the best outcomes you can hope for, I guess."

In theory Dorsey would be richer too, as one of Instagram's earliest investors, but all he felt was sadness. He couldn't stop thinking about Systrom. After all his advice and support, he'd thought they were friends. Why hadn't he called, even just for business reasons? Dorsey had always said the door was open at Twitter. He had always said the price was negotiable. Did Systrom, always preaching about craft and creativity, value Facebook-style world domination more?

As time passed without any explanation from Systrom, Dorsey stopped feeling hurt and started feeling angry. He realized Systrom had never wanted to sell to Twitter. Twitter had been played. Dorsey deleted the Instagram app and stopped posting altogether.

* * * * * * * * * * * *

A few blocks away, around noon, a dozen Instagram employees slipped through their back door and walked down an alley to avoid the press out front. They boarded a shuttle bus that brought them thirty miles south to the vast parking lot encircling Facebook's headquarters, at 1 Hacker Way, Menlo Park.

The buildings were their own industrial island, abutted on one side by an eight-lane highway and on the other by salt marshes at the edge of the San Francisco Bay. Marked by a giant blue thumbs-up "like" sign, the headquarters had so much employee traffic, it was funneled and directed by an army of valets and guards. The weather was about ten degrees warmer than in San Francisco, so the Instagrammers took off their jackets. Before they could see what the insides looked like, they had to give their identification to security contractors to be checked into the system. Security printed out name badges for them, which they were told to wear at all times.

As the Instagram team walked along a carpeted pathway through rows of desks in Building 16, the Facebook employees realized who their guests were. One employee stood up and started clapping, and then the whole room joined in. Many of the Instagrammers, already overwhelmed, became quite uncomfortable.

Their meeting was in the "fish bowl"—the name for the conference room Zuckerberg used, where anyone could guess what was going on simply by looking through the transparent glass. All of the Instagram employees fit inside, some of them on chairs, others piled on a small couch. *That's it,* Facebook employees thought as they walked by. *That's a billion dollars in that room.* The Instagram team looked afraid.

For most of the Instagram employees, it was their first time meeting Zuckerberg, who came off much friendlier than his ruthless, socially inept Hollywood caricature in the 2010 film *The Social Network*. Zuckerberg delivered a more personal version of his blog post, explaining that Instagram had built something important that he intended to keep intact. He also said he intended to welcome all of them into the company. That day, Zuckerberg had posted a picture of Beast on Instagram—his first time sharing on the app in almost a year.

It was slightly reassuring, but none of the other details were hammered out. Employees didn't know when they would officially be part of Facebook, how they would work together, or whether they would make any money off the deal.

They also didn't know whether they would be working out of this corporate playground. On the inside of the circle of Facebook buildings was a wide-open space of asphalt and trees, picnic tables and shops. There was a sushi restaurant, an arcade, a Philz Coffee, even a bank. In the middle of it all was Hacker Square, where Zuckerberg addressed employees for question-and-answer sessions every Friday. The layout, they were told, was inspired in part by the Main Street, U.S.A., zone at Disneyland.

After the tour, everyone was famished. They rode their private shuttle about 15 minutes away from the headquarters, to downtown Palo Alto, which Systrom and Krieger knew well from their Stanford days. With so many people, they ended up at the least Instagrammy location imaginable: the Cheesecake Factory, a chain restaurant with a mix of Victorian, Egyptian, and Roman design themes and a menu so vast it necessitates 21 pages.

The headlines imagined what their day might be like, just based on the numbers.

"The 13 employees of photo-sharing service Instagram are celebrating today after learning they are set to become multi-millionaires," the *Daily Mail* wrote.

"Instagram is now worth $77 million per employee," *The Atlantic* reported.

Business Insider published a list of all the employees they could find, complete with their photos and information scraped off the internet about what schools they went to and where they worked before. Team members were fielding calls and Facebook comments from their friends and family, congratulating them on having made it in life.

But had they made it? They wouldn't have any answers to their money questions for another couple of weeks.

THE SUMMER IN LIMBO

"I write to urge the Commission to open an immediate investigation into whether Facebook has violated the antitrust laws. . . . In hindsight, it is clear that by approving this purchase, the Commission enabled Facebook to swallow up its most significant rival in the social network market."

—U.S. CONGRESSMAN DAVID CICILLINE,
WRITING TO THE FEDERAL TRADE COMMISSION
ABOUT THE INSTAGRAM DEAL IN 2019

All existential questions—about who was going to be a millionaire, about how working at Facebook would change their lives—were put on pause for a weekend of team celebrations in Las Vegas. Kevin Systrom had one rule: No Instagramming. He and Mike Krieger didn't want the media to find out about their trip because they didn't want Facebook to think they'd stopped working hard. Everyone just needed to blow off a little steam.

Most of the trip was covered in full for all Instagram employees, either by the company or through Systrom's connections. One of Systrom's closest friends, the venture capitalist Joshua Kushner, was able to get his firm, Thrive Capital, invested in the latest funding round. Like everyone else, Kushner doubled his money in record time and made a name for himself in the process. So Kushner asked his sister-in-law, Ivanka Trump, to make sure the employees enjoyed themselves. Everyone got to stay in suites in the gold-windowed Trump International Hotel, where they received little personal notes from the heiress, congratulating them.

Out to dinner at the Wynn hotel's steakhouse, Systrom told the team they could order whatever they wanted on his tab, so they asked for caviar and cocktails. The Canadian DJ Joel Thomas Zimmerman, better known as Deadmau5, was playing sets at a nearby nightclub. As he passed by, he recognized Systrom from the news, despite the group's attempt at dining incognito. The DJ congratulated them on the acquisition and lamented that he didn't have the username he wanted. Jessica Zollman set him up with @Deadmau5 right there at the table.

One of the Kushner/Trump associates was in charge of attending to their group. He guided them to a club where they didn't have to wait in line to enter. The waiters delivered alcohol in bottles that shot sparklers out of their tops.

"This isn't very low-profile," Krieger said.

"Every other table is getting the same treatment," a colleague reassured him.

Not five minutes later, the waiters started handing out T-shirts that said "1 BILLION REASONS TO SMILE" on them, complete with the Instagram logo, as well as branded sunglasses. In the moody club lighting, Systrom scrambled to collect them, but the next delivery was even more obvious: an entire cake, emblazoned with "$1 BILLION" in frosting letters.

Mercifully, nobody posted pictures. But plenty were laughing.

· · · · · **·** · · · · · · ·

The trip reinforced the close bonds the group had formed in the intense weeks and months leading up to the deal. There was the time the team got so cold while working through the night that they bundled up in branded swag sent by Snoop Dogg. Or the time they got staff portraits done in old-timey tintype. Or the time they accidentally locked Shayne Sweeney in the building and set off the security alarm. They were a quirky band of twenty-somethings figuring out their lives together, all superfans of the product they were building.

But this was a workplace, not a friend group. And things were getting a lot more complicated.

On the team's return from Vegas, the first news was that Facebook would not be able to help them with resources or infrastructure until the deal was actually approved by regulators. That could take many months, according to Facebook's lawyers. The governments in the U.S. and Europe were investigating whether buying Instagram would give Facebook monopoly powers. Until then, Instagram couldn't be at Facebook's headquarters and wouldn't be able to do much hiring, so they would remain just as overworked.

The second news was personal. Most of the employees weren't getting rich.

A couple weeks after the deal, a Facebook representative came into Instagram's South Park office to join Systrom and Krieger in offering everyone new contracts: new salaries, new stock options, and cash bonuses if they stayed at Facebook for more than a year. One by one, they went into the conference room, and some came out ashen-faced.

Silicon Valley employees often decide to take lower salaries in order to work at a startup like Instagram that offers stock options—the option to purchase shares cheaply at a later date. Those options are restricted based on time. Usually a quarter of the total grant in the job offer becomes available after each year that they continue working, giving an incentive to stick around. For an employee who picks a winning company, the small slice of ownership yields life-changing wealth, like winning the lottery.

Instagram was the biggest mobile app acquisition that had ever happened—the best equity they could have chosen. But if the Instagrammers accepted Facebook job offers, Facebook would cancel their stock options in Instagram and grant them restricted stock units in Facebook instead.

Their equity vesting schedule would start over, as if they hadn't already worked many months.

Only three employees had been at Instagram long enough to have the option to buy a quarter of their Instagram shares and convert them to Facebook shares at a lower price. Everyone else would have no wealth from Instagram stock.

Because Facebook was about to go public, the three long-term employees had to act quickly. At least one of them couldn't actually afford to purchase their Instagram shares to turn them into Facebook ones. Because of the value of the deal, that employee would have needed to get a more than $300,000 loan to afford it. Their lawyer advised against the financial risk, explaining that Facebook was not a safe financial investment to take on as a twenty-something. Nobody knew if the shares were going to do well. (Facebook shares have increased in value by about 10x since Instagram joined, meaning this employee's share would be worth about $3 million today.)

Systrom and Krieger, on the other hand, were awarded life-changing sums. Krieger solidly owned 10 percent and Systrom 40 percent, and so netted an estimated $100 million and $400 million, respectively, per the original deal price. Systrom was proud; he told friends that the day after the deal, he went into the local deli to buy five copies of the *New York Times* and was amused that the cashier didn't recognize him as the man pictured above the fold.

Both started exploring how to spend this new fortune, in a way that the tight-knit team noticed. Krieger was planning philanthropic efforts, looking into how and where to donate money, and also inquiring about collecting modern art. Systrom started looking for a house and invested in Blue Bottle Coffee. Occasionally Systrom's packages from online shopping would be delivered to the office. Employees noticed his new car, new Rolex watch, and new skis. Money had finally unlocked the opportunity for him to have the best, most finely crafted version of whatever he wanted, like an Instagram feed come to life.

Jessica Zollman, the enthusiastic community evangelist, confronted Systrom about the disparity. He explained that what employees were getting from the deal was nonnegotiable, all hammered out already with Facebook. He tried to make her feel better by saying he'd asked Mark Zuckerberg if it was possible to allow her Pomeranian, Dagger, to come along to Facebook headquarters when the deal closed. Unfortunately, Zuckerberg said Facebook didn't allow dogs. Zollman had been taking Dagger into work and realized that on top of it all, she was going to have to start paying for a dog walker.

Frustrated, Zollman and some of the others took a trip to Santa Monica together without the cofounders to get their minds off it, but found themselves in a sort of group therapy session instead. *If Kevin had just given each of us* one *of his millions,* they'd say to each other, *we wouldn't have to rent apartments anymore. We could invest in startups, or start our own.* They asked their friends and learned that it wasn't uncommon for startup founders to distribute life-changing sums to their employees after a big deal.

In the deal, the founders only had a certain allocation of stock for their staff, and felt that the size of the longest-serving employees' reward shouldn't come at the expense of those who'd worked for less time. They could have avoided a lot of bitterness by offering cash from their own winnings, but didn't think the employees working for such a short time should feel so entitled. It wasn't a Gowalla situation, like Tim Van Damme and Philip McAllister had experienced in the winter—everyone was getting a job with a Facebook-size salary, and everyone would get a bonus of tens of thousands of dollars and more if they worked at Facebook for a year. (When the deal was finally approved, some employees—including Zollman and Van Damme—didn't stay at Facebook long enough to get the bonus. Others, including Amy Cole, McAllister, and Dan Toffey, remain at the company as of this writing.)

Instagram was coming of age in an industry that revered and empowered founders above all else. In the deal contract negotiated with Facebook, Systrom and Krieger are the only two Instagrammers described as "key employees." The magic Facebook was paying for was theirs.

· · · · · · · · · · · · ·

That was only the beginning of Instagram's summer in limbo. For the next few weeks, the headlines were all about Facebook, all the time, because of the company's pending IPO. The social networking giant's stock listing captured the public's imagination after Zuckerberg showed up to meet Wall Street's suited bankers in his usual zip-up hoodie, providing the ultimate symbol of Silicon Valley hubris. The company went public at $38 per share on May 18, giving Facebook a valuation of more than $100 billion, worth more than Disney or McDonald's.

The employees celebrated as Zuckerberg rang the opening bell for the Nasdaq from Facebook's headquarters, in a trading debut riddled with technical errors. And then, the next day, the stock started falling. Investors realized that the company didn't yet make money off mobile advertising, even as their users were abandoning desktop computers and spending more and more time on phones.

Shareholders brought a class-action lawsuit, claiming Facebook had intentionally hidden the fact that its sales would slow. Most stock debuts aren't that exciting, but Facebook was a product that 950 million people logged onto every month, some of whom believed in it enough to buy shares. Around the world, members of the social network told stories of investing their life savings into the stock, only to have to pull out after losing so much. This was the stock that paid for part of the Instagram acquisition, so the value of that deal was shrinking now too. The lawyer who had advised against the $300,000 loan was looking smart.

Governments in the U.S. and Europe were starting their investigations into whether the Instagram acquisition should be allowed. Facebook, which in the run-up to the IPO had seemed like it was taking over the world, suddenly had an uncertain future. And Instagram, an 18-month-old app with 13 employees, didn't look very formidable either. The investigations, then, were considered matters of bureaucratic red tape more than matters of public importance. Nobody anticipated how powerful Facebook would become—and how powerful it would make Instagram.

* * * * * * * * * * * *

Antitrust law was not written for modern acquisitions like Instagram. A traditional monopoly was a company with such a hold on its industry that it harmed others by fixing prices or controlling a supply chain. Facebook and Instagram presented no obvious consumer harm because their products were free to use, as long as people were willing to give up their data to the network. Facebook's advertising business was relatively new, especially on mobile phones; Instagram didn't have a business model at all. Something was a monopoly if it undermined its rivals; Instagram had many rivals. Instagram wasn't even the first company to make a mobile photo app with filters.

So the Federal Trade Commission started its investigation with a simpler question. Were Facebook and Instagram competing *with each other*? If they were, it would reduce competition in the marketplace if they were allowed to merge.

First, regulators needed a clear picture of what Instagram thought of Facebook and vice versa, based on internal emails and text messages. Oddly, the FTC would not be gathering this documentation itself. The lawyers for Facebook and Instagram—the same ones who had worked on the deal—were now tasked with finding any evidence showing that the deal shouldn't go through. They were paid by the companies to investigate the companies.

Employees surmised the federal government didn't have the resources to do its own digging. They were shocked to learn the scenario was routine for deal approval in the U.S. Despite the obvious conflict of interest, the lawyers had an incentive for doing a thorough job—the threat of being disbarred if they did not. Instagram's lawyers at Orrick, Herrington & Sutcliffe asked the founders and some of the longest-standing employees to turn over all their email and text histories. They even pored through Systrom's written notebook, page by page, seizing on items that the FTC might find problematic.

At one point, they found a concerning text message—about the expensive bourbon Systrom had gifted to his employees when Instagram passed Facebook in app store popularity. The Orrick lawyers asked Shayne Sweeney what it meant. He told them that Facebook was one of the most popular apps in the world, and that beating them would be a meaningful milestone for any startup, not just for a competitor. He never heard whether that was a satisfying answer.

The law firm Fenwick & West was conducting a similar probe on the Facebook side. After the lawyers presented their materials to the FTC, Systrom and Zuckerberg were asked to go to Washington, DC, for further questioning. Zuckerberg declined the invitation, choosing to do the interview over video conference. But Systrom went, and sat through gentle interrogation by a room of junior employees, some of whom were clearly excited to meet the head of Instagram. They asked him a lot of technical questions about how Instagram worked, perhaps trying to suss out whether Facebook was telling the truth that Instagram served a completely different purpose in consumers' lives than Facebook did.

In information it gave to another regulator, the U.K. Office of Fair Trading, Facebook made the case that while it wasn't directly competitive with Instagram, its just-launched Instagram copycat app called Facebook Camera was. Other apps, like Camera Awesome and Hipstamatic, were downloaded three times more than Facebook Camera, and Instagram was downloaded 40 times more. The argument smartly reframed Facebook as an underdog, trying to compete in a tough new market, as opposed to a giant with 950 million users.

The market sounded crowded the way Facebook described it. The company said there were plenty of other apps like Instagram, including Path, Flickr, Camera+, and Pixable. So the U.K. regulators said they were convinced that allowing the acquisition wouldn't remove competition from the market. The Office of Fair Trading wrote in its report that it had "no reason to believe that Instagram would be uniquely placed to compete against Facebook, either as a potential social network or as a provider of advertising space."

They didn't realize Instagram had already won. The only names on the list that were truly similar to Instagram, complete with filters and social features, were Path, which had fewer than 3 million users, and Hipstamatic, which had peaked at 4 million users and was about to lay off half a dozen of its employees. PicPlz, the app that Systrom and Krieger were so determined to beat after Andreessen Horowitz's investment in 2010, had shut down in July 2012 and wasn't even mentioned.

The regulators were shortsightedly looking at the current market-place and ignoring what Facebook and Instagram had the potential to be in a few years or even months.

The real value of Facebook and Instagram was in their network effects—the momentum they gained as more people joined. Even if someone enjoyed using an Instagram competitor like Path more, if their friends weren't on it, they wouldn't stay. (Path shut down in 2018 after selling to a South Korean company, Daum Kakao, three years before.) Zuckerberg understood that the hardest part of creating a business would be creating a new habit for users and a group they all wanted to spend time with. Instagram was easier to buy than to build because once a network takes off, there are few reasons to join a smaller one. It becomes part of the infrastructure of society.

That's why Zuckerberg was ignoring the headlines that called the $1 billion price ridiculous, and was unconcerned that Instagram had no business model. Making money, in Zuckerberg's opinion, is something to try only once a network is strong enough, so valuable to its users that advertisements or other efforts aren't going to turn them off. Facebook's users were comfortable with sharing their intimate data on the social net-work before they had any reason to question the site's motives.

The network effect was also why Facebook would eventually recover from its investors' panic about making money on mobile. Facebook had millions of users on its mobile phone app—it just hadn't fully turned on the money machine. Instagram's network would be lucrative one day too. The way Zuckerberg saw it, as long as there were users, there was a potential to create a business around them—and the more users, the better.

Instagram was also a threat to the thing Facebook wanted from its users the most: time on its site. Facebook was in fierce competition with any other network that people would choose to visit in a spare moment—anything that allowed people to see what was going on in other people's lives and post about their own. The stronger Instagram's network got, the more it would become an alternative to Facebook for those moments of blank space in a day—in a cab, in line for coffee, bored at work.

Facebook was a master at strategically massaging the truth to reduce government scrutiny, presenting itself as a scrappy upstart when it wasn't. But the company's paranoia was real. Any fast-growing social media product was a threat to Facebook's network effect and the time users spent there. It was Facebook's job to not let anyone else catch up; Zuckerberg had instilled this value in his employees by ending all staff meetings with an unambiguous rallying cry: "Domination!"

There were signs Instagram was achieving a winner-take-all effect. Its growth was accelerating. At the time of the acquisition, the company had 30 million users. By the middle of the summer, it had more than 50 million.

The Office of Fair Trading's report says nothing about network effects, indicating that Facebook didn't fully explain its logic behind the deal. They took an opposite read on Instagram's growth. "Whilst this indicates the strength of Instagram's product, it also indicates that barriers to expansion are relatively low and that the attractiveness of apps can be 'faddish,'" the report said.

Today, Facebook is still the most dominant social network in the world, with more than 2.8 billion users across several social and messaging apps, and the primary driver of its revenue growth is Instagram. Analysts would later say that approving the acquisition was the greatest regulatory failure of the decade. Even Chris Hughes, one of the cofounders of Facebook, would in 2019 call for the deal to be undone. "Mark's power is unprecedented and un-American," he wrote in the *New York Times*.

The FTC's investigation in the summer of 2012 happened behind closed doors, with no public report about its findings. Facebook says "the process was both robust and thorough," led by "very competent staff." When the proceedings closed, the regulator sent letters to Facebook and Instagram telling them that "no further action is warranted at this time." The letters included a caveat that they might take another look later, "as the public interest may require."

· · · · · · · · · · · ·

Instagram needed to sell to Facebook because Systrom and Krieger had been slow in hiring. They were so particular about picking employees who would be a perfect fit, despite being so frenzied with keeping the site alive. Once they turned down Twitter and raised the $50 million in venture capital from Sequoia, they were still, in the words of one investor, *too hungry to eat*. They probably needed ten times as many employees in order to grow fast enough to give those investors the hefty return on investment they expected.

They were exhausted. Selling was the simplest way to solve the problem. Facebook had more than 3,000 employees—some of the smartest engineers in the world. Once Instagram joined them, if the FTC allowed, they'd be able to recruit from within. They would just have no relief in the meantime. Their investments in employees and infrastructure were on hold, pending the deal close, at a time sign-ups were accelerating. As usual, Krieger's erratic sleep schedule was the surest sign of Instagram's unending expansion.

One Friday night at the end of June, Krieger was in a cab to dinner with his girlfriend of two years (and now wife), Kaitlyn Trigger, on a rare weekend trip to Portland, Oregon, to explore the restaurant scene. The familiar notifications about issues with the app started, but he figured Sweeney would be able to handle it, with the help of a more recent employee, Rick Branson.

Unfortunately, it was not a usual outage. The whole internet was down. Or at least, the Amazon-supported internet. Every company that had built its servers with the support of Amazon's infrastructure—including Pinterest, Netflix, and Instagram—was completely offline due to a storm on the East Coast. Most of those companies had many dozens of back-end engineers at the ready to fix such problems. Instagram had only three, one of whom had been on the team just two weeks.

"We need to turn the cab around," Krieger instructed the driver, before saying sorry to Trigger, who was used to such crises.

Sweeney got the alert during a San Francisco Giants baseball game at the stadium with his family, who was in town for a reunion. He apologized to his relatives, left the stadium during the third inning, and walked the few blocks to the South Park office.

When the servers came back on, all of Instagram's code needed to be rebuilt from scratch. The data still existed in full, but the computers needed to be retaught what to do with it. Krieger and Sweeney spent the next 36 hours patching it back together, Branson pitching in wherever he could, but feeling useless since he was wasn't familiar with the code-base yet.

It was the most dire server problem in company history. Instagram was now important enough to be mentioned in every press story about the meltdown, alongside Pinterest and Netflix. Coworkers, none of whom did that kind of engineering, sent ice cream to the office as support. Sweeney ate several scoops to try to make it through the night, though he accidentally fell asleep multiple times on his keyboard.

The infrastructure wasn't the only problem bubbling up to an intensity the tiny team could barely handle. Spam was everywhere on Instagram. So was troubling and abusive user content, which the community team could no longer finish sifting through in its shifts—and which was starting to appear in their nightmares. Frustration over the financials aside, selling to Facebook might give employees their lives back.

• • • • • • • • • • • • •

Facebook in its discussions with regulators was right about one thing: Instagram was reaching a different audience than they were. Facebook required real names; Instagram allowed anonymity. Facebook had re-sharing and hyperlinks; Instagram did not. Facebook was about mutual friendships; on Instagram you could follow people even if they didn't follow you back.

Facebook was like a constant high school reunion, with everyone catching up their acquaintances on the life milestones that had happened since they'd last talked. Instagram was like a constant first date, with everyone putting the best version of their lives on display.

On Instagram, people wanted to post things that would attract the adoration of an audience. If an image was beautiful, well designed, or inspirational, it would do well on the app. So people changed their behavior, seeking out more things that would do well, appreciating well-plated meals, street-style fashion, and travel. Phrases like "outfit of the day" and "food porn" and "Instagrammable" entered the vernacular as the company grew. Nobody said "Facebookable." Instagram had a higher bar.

Systrom, who sought out well-crafted things and experiences in his own life, did want images on Instagram to meet a higher bar. But, he'd say, the pressure wasn't on the users, it was on Instagram—to deliver a quality boost automatically with its filters.

It was also Instagram's job to define itself, in a way that didn't lean on what was popular. Instagram did have a "Popular" page, which was the only place content was sorted by a computer the way the Facebook news feed was. "If you go on the 'Popular' page it's pretty obvious that it is boobs and dogs and really sexy girls that still drive it," explained Jamie Oliver, a celebrity chef who otherwise adored Instagram, at a conference in 2012. But in the community team's eyes, Instagram wasn't about that. It was a world of interesting niches, cultivated through the team's training blogs and InstaMeets, and curated so carefully through its suggested user list.

Instagram was teaching and rewarding storytelling. The app was "more valuable the better the stories are," according to Bailey Richardson, the keeper of the list.

Facebook avoided human decision-making around what people saw in their feeds; Instagram loved picking favorites. Everyone they picked would instantly get a bigger audience, becoming a model citizen for others on the app, so the choices were critical. In an ideal world, the suggested user list would be full of people like Drew Kelly.

They'd discovered his account that summer. Instagram was designing a product to put users' photos on a map—their distraction from the post-deal malaise—and the community team noticed someone using Instagram in the last place they expected: North Korea. The account's keeper was Kelly, an expat teaching in Pyongyang. Kelly saw an opportunity to depict a different side of North Korea, beyond its oppressive government. He tried to chronicle stories of students facing exams, conversations in a cafeteria, walks through the local market.

"My room was bugged, my phone was tapped, my Instagram was monitored, and every conversation I would have with a local was reported and sent to the Ministry of Foreign Affairs," he remembers. He still managed to post via spotty Wi-Fi from a school that provided a rare connection to the outside world, representing two-thirds of the country's bandwidth. In Kelly's world, Instagram was a tool for micro-diplomacy, as he called it. A bridge for human understanding.

If Kelly was representative of the user base, Instagram had graduated beyond the frivolity of latte art and into matters of serious world importance, which Twitter's Ev Williams had never thought it would reach.

Kelly was featured on Instagram's blog but declined placement on the list, fearing for his security. He knew the list's power. The community team still tried to use its influence for good, elevating photographers and bakers and artisans who once they were popular enough were able to quit their day jobs and pursue their passions full-time.

But to their dismay, several of the people who gained followers from the list—the first regular people to taste Instagram fame—decided to capitalize on the opportunity.

Liz Eswein, who ran the @newyorkcity account, now with almost 200,000 followers, had friends in the media and advertising industry who were paying for placement in magazines with a smaller audience than she had. Still recovering from Lyme disease, she tried to make money by selling access to them. Nike paid her a nominal sum—less than $100—to post a blurry picture of disabled endurance athlete Jason Lester, tagging @nike and using the hashtag #betterworld. She worked with two other Instagrammers to start a tiny advertising agency off the idea. Their first client was Samsung, which asked them to shoot for Instagram with the Samsung Galaxy Note, using the hashtag #benoteworthy. Soon, other Instagrammers with large followings started to follow her example.

The Instagram team didn't think that the accounts it promoted should be profiting off their followers' attention, especially if they were meant to be the models for everybody else. So that summer, Instagram culled its suggested user list from 200 accounts down to 72, in an attempt to quash some of the brand activity. In an email to the members of this suggested user list, the company explained their reasoning: "While we're excited that people have a large enough audience to start experimenting with [advertising], it's not the type of content we envision being the right experience for new users."

Instagram was not supposed to be about obvious self-promotion, Systrom said. It was about creativity, design, and experiences. And honesty. "I think what makes it so good is the honesty that comes with the photos," Systrom said at LeWeb, the French tech conference, in June 2012. "The companies and brands that use Instagram—the best and most successful ones—are the ones where it comes across as honest and genuine."

The word choice was telling: "where it *comes across* as honest and genuine." It's not that Systrom was against people selling products on Instagram. He just wanted them to do it in a way that masked their financial incentives.

Systrom didn't want Instagram to turn into a collection of unsightly roadside billboards. When users posted about brands, instead of being so obvious, it would be best if they acted like they were letting their audience in on a life secret, or if they put the product in a spread of other beautiful things, or if they told a story.

Years later, the Insta-famous hawking products on the app would not be called "salespeople" or "celebrity endorsers." They would be called "*influencers*." Appearing genuine would be a top priority. But actual honesty would be difficult with so much money on the line.

And Systrom, now the head curator of a visual revolution, seeing his product spark changes in human behavior around the world, cavorting with celebrities and learning to believe in his own taste and vision, was going to have to fight for the kind of network he wanted to build—not just with Instagram's new creative class, but with Facebook, the most utilitarian company in Silicon Valley.

• • • • • • • • • • • •

Jack Dorsey, meanwhile, was grappling with the fact that he'd essentially been the first Instagram influencer, selling Instagram itself. And now the shares he owned in the startup, tainted by Systrom's betrayal, were turning into something worse: shares in Twitter's enemy no. 1.

Before the deal, it had seemed possible that Twitter would be bigger than Facebook one day. Now he wasn't so sure. It made him anxious, being a Twitter board member and owning Facebook, like he was cheating on a lover. Dorsey needed to wait a few months to dump the stock because of legal restrictions on insiders selling after an IPO. He started looking for other companies that could fill what he now saw as a void in Twitter's product, which needed more visual storytelling.

Twitter's leaders decided that since Instagram was going to be a part of Facebook, it should be treated like a massive competitor—not a scrappy startup. So later that summer, an error message started to appear for Instagram users. They could no longer use the list of people they followed on Twitter to find their friends on Instagram. Twitter's engineers had blocked Instagram from access to their network.

Twitter confirmed that it was no longer helping Instagram grow. "We understand that there's great value associated with Twitter's follow graph data, and we can confirm that it is no longer available within Instagram," spokeswoman Carolyn Penner told *Mashable*.

It was a salty end to the alliance, but Systrom could feel no ill will. The Twitter executives never had a chance to counteroffer. Legally, per the terms of the Facebook deal, Systrom wasn't supposed to give them one.

· · · · · ● · · · · · ·

Once federal regulators allowed the acquisition to go forward, only state regulators stood in the way. On a brisk Wednesday morning in late August, a couple dozen men in suits congregated in a windowed conference room on the sixth floor office for the California Department of Corporations in San Francisco. Systrom, the tallest of them, elevated his business look with a tie clip—a detail the press mentioned because of Zuckerberg's pre-IPO sartorial scandal. The tables had been joined to form a large rectangle, Facebook lawyers and Amin Zoufonoun on one side, Instagram lawyers and Systrom on the other.

Zuckerberg was not present, as he was not required to be, and as this was not expected to be difficult. But it was rare: an open interrogation about all the decisions made behind closed doors. Members of the press and public were allowed to attend and listen over the phone.

It was called a "fairness hearing"—a rarely used option California offers to companies with uncomplicated deals, so that they can issue stock with the approval of state regulators instead of going through the longer federal process. The Department of Corporations intended to question both parties to ensure that the transaction would be fair to all 19 shareholders of Instagram.

Zoufonoun acknowledged that the deal had come together very fast—without any financial advisors or investment banks. But he emphasized that all of the terms had been negotiated extensively. (As extensively as they could be over beers on Easter weekend.)

When it was Systrom's turn under oath, he started by defining his company. He said Instagram "allows people to share their photos in a fast, beautiful, and creative way to multiple different services all at once, including Instagram's own proprietary network." He explained that Instagram, after about two years in existence, operated at a net loss of $2.7 million, and had $5 million cash in the bank, with 80 million registered users.

"How does Instagram create revenue?" the California acting commissioner Rafael Lirag asked.

"That's a great question," Systrom said. "As of right now, we do not." Without the acquisition, he explained, they could probably continue operating on their own, though he couldn't say for how long. But with Facebook, the Instagram shareholders had a much more secure future.

Lirag pushed Systrom on whether Instagram might be better off on its own, or selling to a different company. Facebook stock that day was trading at $19.19. Had he ever expected it would fall like that, thereby pushing the value of the deal below $1 billion?

"In a large part, the billion-dollar valuation headline was really something generated from the press," Systrom said. (Of course, it wasn't. The number was in Facebook's public commentary on the deal, and also what Systrom told his employees.)

But there were no other offers? Ivan Griswold, a lawyer from the department, wanted to know.

"No, we never received any offers," Systrom said. "We have, throughout the course of business at Instagram, talked to other parties, but we never received any formal offers from anybody else."

"Immediately prior to the negotiations, did you receive any offers from—"

"We never received any formal offers or term sheets, no," Systrom cut Griswold off. His interruption signaled his discomfort with the question. He might not have been serious about Twitter, but Twitter had been serious about acquiring Instagram.

As a last step, department representatives asked if there were any questions or concerns about the deal, from people in the room or on the phone lines. If Twitter wanted to protest, this was the Hollywood wedding moment to speak now or forever hold their peace. But they weren't there.

"The terms and conditions of the proposed transaction have been found to be fair, just, and equitable," the commissioner concluded. The hearing was over, one hour and 22 minutes after it started. In about ten business days, Facebook could issue shares to buy Instagram, and Instagram employees would become Facebook employees.

• • • • • • • • • • • •

Executives at Twitter who had tried to buy Instagram would later argue—anonymously, on page B1 of the *New York Times*—that Systrom had committed perjury. But only after Facebook did something to piss them off.

The press was the only leverage Twitter had now. The deal, which had sailed through approvals in six months without much conflict or delay, would weaken Twitter's promise while affording Instagram all the competitive advantages of the biggest network in the world. And it would eventually ensure that the main alternative to Facebook was a product also owned by Facebook.

MOVE FAST AND BREAK THINGS

The Monday after the deal was finalized, Instagram employees hopped on their Wi-Fi-equipped Facebook buses in a forced embrace of their new one-hour commute. When they arrived, they got their employee badges and desk assignments in a new space, behind a glass garage door with blue trim.

The new Instagram headquarters was smack-dab in the middle of the Facebook office park, which employees called a "campus," as if everyone were still in college. Right outside, the word "HACK" was painted on the cement in gray letters so big, passengers flying into San Francisco International Airport could see it from their planes. Instagrammers would be working next to rustic outdoor fire pits and the Sweet Stop, a shop that provided anyone with free cupcakes and soft-serve ice cream.

Systrom was coming to terms with the practical realities of the deal's close. His people, so focused, fast, and passionate, were about to be part of a massive corporation, with all the comforts that entailed in the age of the Silicon Valley talent wars. Free food, free transportation to and from work, free sweatshirts, water bottles, and parties. What if they lost their drive? What if they felt like they'd *made it* and stopped working as hard?

Most outsiders assumed Systrom's own journey was over. In Silicon Valley, it was common for founders, once their companies were acquired, to "rest and vest"—spend the next four years at the new parent company waiting for their stock options to vest and make them millionaires, without having to do much work. So Systrom would get annoying questions about what he'd been up to since the deal. *Are you kidding me?* he'd think. *I'm still building this thing.*

Systrom posted a picture on Instagram of his still-small 17-person team in front of the garage door: "First day in the new offices! Can't wait to show everyone what's next!" Later that night, he decided to use one of the fire pits. It was only about 6:30 p.m. but disturbingly, no Facebook employees remained. "Heading home after a great first day," he posted, with a picture of the fire.

That week, as if to stoke his worries further, Facebook threw a mid-day party. They were celebrating the fact they had reached 1 billion active users around the world—a milestone that no social network had ever accomplished. Employees lapped up the free-flowing booze in a scene that harkened back to Facebook's fraternity-style early days, when they were in their suburban Palo Alto pool house rife with beer pong.

A couple of Instagram's designers, welcoming the break after the exhausting summer, participated in the festivities and returned to the garage tipsy. Systrom was dismayed. "*We* didn't hit a billion users," he said. Time to get back to work.

· · · · · • · · · · · ·

Systrom and Krieger said yes to Facebook so that Instagram would be big, powerful, and important one day. There was an obvious way to make it so—to simply do as Facebook did. But with the promise of independence, they still wanted to be visionaries, asserting Instagram's role as a startup within a big company, with a different brand and ethos.

They would only fit into their new home if they learned to adhere to a corporate philosophy more attuned to metrics than to cultural moments. Facebook wanted metrics—milestones like 1 billion users—so it could swallow from an even bigger fire hose of data on human interactions. The data could help improve the product so people would spend more time there, creating even more data through their posts and comments. Then the data would enable Facebook to sort people into smaller audiences that advertisers would want to sell to.

If Facebook's employees were overindulging in the middle of the week, it was because they needed the boost. Employee morale tracked closely with the stock price. Facebook shares, which had started trading at $38 in May, had lost about half their value by that September, and Zuckerberg was on a warpath to turn things around. He would refuse to give feedback on products unless they were designed first for mobile phones, so the company could catch up to the rest of the industry, including upstarts like Instagram.

The Instagram deal had been approved near a historic low for the stock. The final cash and stock price Facebook recorded for Instagram is $715 million—not the $1 billion number that made all the headlines. Still, the billion-dollar number was what made Systrom and Krieger feel like they'd come into the company with something to prove.

They could feel the skepticism. Besides the public commentary from friends and media, Facebook employees were openly questioning their managers about the value of the deal, looking into the glass garage as they walked by, to try to understand it. If this was what it took to get rich, they'd say, maybe they should just quit and build a competitor, in the hopes that Facebook would acquire their company.

Nothing Instagram-related was on Facebook's strategic road map for the second half of the year. Even though it was a mobile-only product, it didn't make any money and, in Facebook's opinion, wasn't big enough to start.

It was also quite possible that from Facebook's perspective, Instagram was still a threat.

• • • • • • • • • • • • •

Facebook's users had been addicted to posting every single photo from every single party and vacation, and tagging their friends—which in turn caused all those friends to get emails and little red notification dots luring them back to Facebook. Every visit mattered to the business. But based on digging into recent data, Facebook saw that kind of photo-sharing behavior was in the early stages of decline, and thought perhaps Instagram could be to blame.

Gregor Hochmuth, the Instagram engineer, got an invite to a lunch meeting with the Facebook Camera team—the group that had launched an Instagram copycat app, oddly the month after the Instagram acquisition. "Our job was to kill you guys," they explained to Hochmuth over lunch. At the time, Facebook couldn't be certain the acquisition would close. And Hochmuth wasn't sure how to read their tone, or how he felt being their colleague.

Soon after, Instagram employees were invited to a meeting with Facebook's all-star growth team. Their message was clear: Instagram wouldn't get any help adding users unless they could determine, through data, that the product wasn't competitive with Facebook.

The Facebook growth team built upon Hochmuth's rudimentary analytics, trying to understand what kinds of people were joining Instagram, and whether having that app meant sharing fewer photos on Facebook. Instagram had been under Facebook's umbrella for just a few days, and already the bigger company was willing to let it languish if there was a chance it could threaten the main product.

Ultimately, the team's study was inconclusive, and Instagram was allowed access to Facebook's growth expertise. The whole ordeal seemed like overkill, as Instagram only had 80 million users, compared to Facebook's 1 billion. But it also served as a lesson in what had made Facebook so successful in the first place.

· · · · · · · · · · · ·

Facebook's overarching goal was to "connect the world" through social networking. The language in marketing materials sounded noble, like Facebook was in the business of enabling empathy for humankind. In practice, the effort was quite literal: to get as many people as possible to use Facebook as often as possible. Every single activity at the company— deciding what new features to build, how to design them, where to put them in the app, how to push them to users—stemmed from a religious obsession with growth, marketed to employees as a moral mission.

While Instagram was trying to give people new interests, Facebook was using data to figure out exactly what people already wanted, and then giving more of it to them. Whatever Facebook observed in activity from its users, it could use to define their likes and dislikes numerically, and then adjust those measurements if needed.

Facebook automatically cataloged every tiny action from its users, not just their comments and clicks but the words they typed and did not send, the posts they hovered over while scrolling and did not click, and the people's names they searched and did not befriend. They could use that data, for instance, to figure out who your closest friends were, defining the strength of the relationship with a constantly changing number between 0 and 1 they called a "friend coefficient." The people rated closest to 1 would always be at the top of your news feed.

Facebook was all about personalization, not just for the ordering of its news feed but for advertiser targeting. A business could sell something with a message tailored to Facebook's cat lovers in Toronto with college degrees, and sell the same product differently to Facebook's blue-collar dog lovers in Vancouver. It was a revolutionary advertising business, because on television advertisers had no idea who they were reaching.

But in order to get that data, Facebook had to grow. They needed to grow not just in number of users, but in time spent by those people, taking all those little actions that added up to vast stores of knowledge about what people wanted—in their news feeds, in their advertisements, and in their Facebook product. And the more people who joined the product, and the more content they produced, the more slots there would be in the news feed for brands to place ads.

The growth team, led by Javier Olivan, was also able to quickly detect, diagnose, and fix problems. He and his team tracked user behavior on massive computer monitors, with charts that were segmented out by type of activity, country, device, and more. If something went wrong— say, the growth rate all of a sudden slowed in France—someone would investigate and find out that the Facebook contacts importer for a popular French email system had broken. They would fix that, then move on to the next snafu, and the next.

Everyone at the company had access to the whole Facebook code base and was allowed to make changes to the product without much oversight. All they needed to prove was that their edit caused a boost, however small, for some important metric, like time spent on the app. That allowed engineers and designers to work a lot faster, because there was less arguing about why or whether they should build something. Everyone knew that their next raise would hinge on whether they affected growth and sharing. They weren't held accountable for much else.

Threats to and opportunities for Facebook's product were evaluated with the same depth of analysis as everything else. Facebook had access to data that tracked how often people were using different apps on their smartphones. The data acted as an early warning system for a potential competitor's rise. If there was any chance Facebook could build its own version of the app that might ultimately reach more people, they would try, immediately. If it didn't work, that's where acquisitions like Instagram came in handy.

A few years later, as Facebook's power grew, its tactics for detecting and paralyzing competitors would come under intense scrutiny. Facebook's strategy for giving people what they wanted would be accused of addicting the world to the digital equivalent of junk food. Its data collection would spark further panic over privacy. But for now, with the stock down in an era before the public reckoning, Facebook was singularly focused on demonstrating that it could create a viable long-term business, even on mobile phones, proving all the haters wrong.

"This Journey Is Only 1% Finished," the posters around campus declared.

"The Riskiest Thing Is to Take No Risks."

"Done Is Better than Perfect."

"Move Fast and Break Things."

Employees rarely challenged these assumptions. They provided a comforting clarity about what success looked like, all outlined in that helpful little book from employee orientation. "It would be easy to get complacent and think we've won every time we bring ourselves to a new level, but all that does is just decrease the chance we'll get to the next level after that," Zuckerberg wrote in an email in 2009, which is memorialized in the handbook. Facebook was forever the underdog, no matter how big it got.

The Instagram team was too small to have codified what their values were, but now, confronted with Facebook's hacker culture, they knew what they *weren't*. Instagram wanted things to be carefully considered and designed before they were released to people. Humans, not numbers. Artists, photographers, and designers, not DAUs, the Facebook term for "daily active users." They didn't want to limit people to their likes and dislikes; they wanted to introduce them to things they'd never seen before.

Regardless, Instagram had to figure out what their metrics were at this point. Facebook's growth team told them not to be naive. One day, Instagram's growth would inevitably slow, and they would have to understand what enticed their users to spend more time on the app, and what barriers prevented them from coming back. *You can thank us later,* the growth gurus told them.

That threat seemed like such a far-off possibility. Instagram's app was still adding users so fast that employees could barely keep it online. Instagram was told that the recipe for growth at Facebook—sending notifications and reminder emails, clearing sign-up hurdles, understanding the data, playing defense—was the most important thing to learn if they wanted the app to be truly important one day. It was also the thing that, if implemented badly, could completely kill the good vibes Instagram had going with its community.

Facebook users were already used to the company pushing the boundaries of privacy and comfort to accomplish more sharing on its products, and then apologizing if it didn't work out. One of the earliest examples was in 2006, when the company moved personal Facebook page posts into a public "news feed" overnight without warning, causing a dramatic uproar that eventually subsided when everyone became addicted to the new feature.

Over the years, Facebook had learned that people would get mad about breaches of privacy and then forget about them because they actually enjoyed what they were seeing—after all, users were getting exactly what Facebook thought they wanted, based on their previous behavior. Usually, people calmed down. And if they didn't stop being angry, Facebook could reverse its decisions or come up with a version of the product people weren't as mad about. *The riskiest thing is to take no risks.* The only real consequence, so far, had been a settlement with the U.S. Federal Trade Commission, which said the company had to get users' express consent before siphoning off a new kind of data.

Instagram employees had no desire to warp their brand into Facebook's. But they lacked a way to explain the value of their good reputation in numbers Facebook could understand. In turn, Instagram's precious sensibilities became the butt of jokes at Facebook. They simply took themselves too seriously—and Systrom wasn't helping.

A few weeks after the deal closed, he joined Facebook executives for a meeting with a couple of the company's top advertisers, at Evvia Estiatorio, a Greek restaurant in Palo Alto. Before it started, he ran into the advertising vice president Andrew Bosworth, a tall bald man who was one of Zuckerberg's top lieutenants, known to speak his mind. Bosworth wore a T-shirt that said "Keep Calm and Hack On."

"I like your shirt," Systrom said.

"Thanks, I got it at a hackathon in London," replied Bosworth, who went as Boz.

"Oh, I thought it said 'Keep Calm and *Rock* On.' I actually don't like that shirt," Systrom replied. *Ugh, hackers.*

"Hey, man, I hear you, but at least my shirt fits me," Boz said. Systrom's appeared too tight.

"This shirt costs more than your car," Systrom retorted, ready to fight in defense of fashion as art, before onlookers dragged the men into the meeting, Boz rolling his eyes, thinking Systrom was either arrogant or insecure or both. The shirt was from Gant, a men's boutique for yuppies. Boz drove the 10-year-old Honda Accord parked outside.

· · · · · **·** · · · · · ·

Exactly what kind of authority Systrom and Krieger had at Facebook was unclear. They came in with regular employee ranks, as a product manager and an engineer, respectively. Systrom reported to Mike Schroepfer, recently promoted to chief technology officer, while his transition was managed by Dan Rose, Facebook's head of business development. Neither of those managers asked too much of Instagram, at Zuckerberg's behest. He'd told the entire company to not bother the tiny team and let them do what they did best.

But Zuckerberg did have some opinions. Besides sending in the growth team to investigate how big of a threat Instagram posed to Facebook photo sharing, his first ask for Instagram was to allow people to tag each other in photos.

At Facebook, product requests were ranked by priority number, with ones and zeroes being top priority. The only thing above that priority level, superseding anything else on the road map, was unofficially called a "ZuckPri," which meant that Zuckerberg was tracking the progress. Photo tagging on Instagram was a ZuckPri. It had been such a boost for Facebook in its early days, he was sure that it would work for Instagram.

Systrom wanted to prioritize photo tagging, too—but subtly, not in the way Facebook expected. Systrom and Krieger balked at the idea of sending their users emails about whether they'd been tagged in something, or emails at all. They didn't want to be annoying or trade the trust they had gained with their community for a temporary boost. They also didn't think the activity merited sending anyone a push notification, which would then produce a red badge on users' phones that they'd have to clear. If Instagram used notifications too much, they would become meaningless, the founders argued.

That was the benefit of being smaller. At Facebook, the news feed was full of competing features. Every product manager working on every aspect of the social network—events, groups, friend requests, comments—wanted their team's tool to be granted an opportunity for a red dot, or a push notification, so that they would get a fair shake at meeting their growth goals and getting a good performance review. The idea that one might *not* add a notification with a new feature was a foreign concept—Facebook championed growth at all costs.

Instagram got its way, because Zuckerberg had insisted on allowing the division to think independently. As a result, when Instagram introduced photo tagging, it did nothing to boost growth. But using the app remained a pleasant experience, whatever that was worth. And people could now see a helpful record of the pictures they were in beyond their own feeds.

Krieger and Systrom started to understand the strengths of their position: they could learn all of Facebook's tricks, and then they could understand the pros and cons of those moves by looking at how Facebook's own product had succeeded or failed. Then, hopefully, they could decide to take a different path if they thought it necessary.

• • • • • • • • • • • •

For the most part, Zuckerberg had told all his employees to leave Instagram alone, except when they needed help. Since it was his first time acquiring a company that he intended to keep intact, he didn't want to screw it up by being overly prescriptive. He was waiting for the network to get stronger and give Instagram staying power first—just like he waited to put advertising on Facebook until users had built a habit there.

But Instagram had never been part of a big company either, and so it took them a while to understand how to ask Facebook for resources. Because Instagram didn't have the engineering power to build systems as vast as Facebook's, they'd invented things with more of a personal touch. But their way of operating was becoming unwieldy as millions more users joined each month. Systrom and Krieger didn't want unnecessary push notifications, but they were willing to make trade-offs on quality in some other areas to help the app get bigger faster.

Facebook's resources helped relieve burdens on employees like Jessica Zollman. Zollman, the Instagrammer who had worked on the earliest community moderation tools and had become so familiar with the threats to its users, was sure she wouldn't be able to find and solve as many problems as Facebook's vast army of contractors could.

To better serve the millions of people joining Instagram, she worked on transitioning content moderation, so that whenever people clicked to report something awful they saw on Instagram, it would just be funneled into the same system of people who were cleaning up Facebook.

Facebook had low-wage outside contractors quickly clicking through posts containing or related to nudity, violence, abuse, identity theft, and more to determine whether anything violated the rules and needed to be taken down. Instagram employees would no longer be as close to their worst content. Their nightmares would be officially outsourced.

Facebook could also help Instagram grow in new countries by offering its translation tools. Instagram was already translated into a few languages, with the help of superfans who had volunteered in their countries, but Facebook's system handled many more languages. The decision bothered people like Kohji Matsubayashi, a "language ambassador" in Japan who thought Facebook's version was lower quality.

Matsubayashi had personally, painstakingly translated the Instagram app into Japanese as a labor of love, answering a call posted on Systrom's Instagram. He found that when Instagram replaced his version with the Facebook version, some of the tiny problems he'd solved in the text on the app became problems again. Japanese users were complaining to him about little things, like using the word "写真" for "photos" instead of the more colloquial "フォト."

He wrote an email to Krieger laying out his concerns. "The minor translation things I noticed on 3.4.0 might be the start of losing translation quality and I was afraid, that was why I am writing this message to you," he explained. But there was no response. Facebook's system made sense for Instagram's future, even if the quality was sometimes poorer.

Facebook preached operating "at scale"—serving more users with less employee effort. Handing things off to Facebook seemed to always mean a trade-off, unavoidable if Instagram wanted to grow.

· · · · · · • · · · · · ·

It was also important to Facebook that Instagram grew in ways that served Instagram, not a major competitor. Facebook saw no reason for Instagram photos to continue to display in Twitter posts. The feature that helped the app catch on, through filtered photos displayed by Jack Dorsey, Snoop Dogg, Justin Bieber, and others, was also creating posts, for free, that Twitter could advertise around—not Facebook. Facebook had a new plan—to only display in tweets blue links that would redirect people to an Instagram website where they could see the photo and download the app.

When the change went into effect that December of 2012, the public complained to Twitter, fearing something was broken. But a Facebook spokesperson confirmed to the public that the change was on their end.

The conflict reignited Twitter's sense of unfairness around the deal, and they retaliated by speaking to then–*New York Times* reporter Nick Bilton, who was working on a book about the company at the time. They brought up the summer's hearing, where Systrom had denied getting other offers for an acquisition. Bilton needed proof, so they took him into the Twitter offices, where a lawyer flashed the term sheet Twitter had prepared in March 2012. The *New York Times* lawyers reviewed the story carefully, because it would level a serious accusation: that Systrom had committed perjury.

"Given that the privately traded Twitter is expected to make $1 billion in revenue next year, which would increase its valuation considerably, Instagram investors might have made millions of more dollars," Bilton reported. Nobody knew if Facebook was going to get out of its mobile struggles, but Twitter was on the road to a flashy IPO of its own.

Mark Leyes, a spokesman for the California Department of Corporations, told newspapers that the claim would be considered a "hypothetical situation," not worthy of further investigation unless an "interested party" filed a formal complaint. The definition of "interested party," in this case, was a Facebook or Instagram shareholder. Of course, none of them said anything.

On the Instagram side, only Systrom knew for certain what had happened around the fire pit in Arizona. He stuck by his story. And he told friends that Bilton, a regular attendee at dinner parties with Systrom's founder-and-CEO friends, was only writing the piece because Instagram was important now. Bilton was never invited back to the dinners. And Instagram pictures never displayed in tweets again.

• • • • • • • • • • • • •

Later that December, Instagram, usually a media darling, faced another press crisis. There were no lawyers among the early employees, so when the startup created its first "terms of service," they simply copied and pasted some boilerplate language from the internet, and then edited it to be Instagram-specific until it looked right. As a public company, Facebook had standards that were a tad higher. In December, Instagram accepted their counsel's edits adjusting the language for the new era, and for a future that might include making money and sharing information with Facebook.

Systrom and Krieger didn't read the new terms carefully until headlines in the media reacted to them.

"Instagram says it now has the right to sell your photos," *CNET* blared.

"Facebook forces Instagram users to allow it to sell their uploaded photos," a *Guardian* headline warned.

The articles kept coming, advising users that there was no way to opt out of the new rules unless they deleted their Instagram accounts before January, when the terms went into effect. The hashtag #deleteinstagram started trending on Twitter, where people were quoting the following language from the new terms: *You agree that a business may pay Instagram to display your photos in connection with paid or sponsored content or promotions without any compensation to you.*

It certainly sounded like Instagram was going to try to profit off the budding prominence of its photographers and artists. But Krieger and Systrom were just as shocked as the users were. They wanted to open the door to the possibility of advertising, but still didn't have a business model to speak of, especially not one based on selling users' photos.

Mostly, they had completely underestimated how much their users would mistrust—and even hate—Facebook. The angry tweets made it clear the Instagram community was looking for signs that the acquisition had ruined the app forever.

With the internet in a frenzy, Systrom wrote his first-ever Zuckerberg-style apology blog. The post explained that the language was unintentionally confusing and would be removed.

"Instagram users own their content and Instagram does not claim any ownership rights over your photos," Systrom said. "We respect that there are creative artists and hobbyists alike that pour their heart into creating beautiful photos, and we respect that your photos are your photos."

As he clicked publish, Systrom was watching a chart—one of the new analytical tools from the growth team—that showed how Instagram deletions were climbing. As the public absorbed the news, the deletions stopped, and eventually the app returned to growth.

• • • • • • • • • • • •

Dan Rose, the Facebook executive managing Instagram's integration, watched the ordeal with interest. It proved a few things. First, that Instagram indeed had a very different brand, one that its users cared about deeply. And second, that Facebook would have to be much more careful. Maybe they needed a liaison between the two companies, keeping a closer eye on the differences and figuring out how to deploy resources, translating Instagram's needs into Facebookese.

At the advice of chief operating officer Sheryl Sandberg, Rose called up one of her protégées, Emily White, a rising star in charge of mobile partnerships who had just come back from maternity leave.

"We're really screwing this up," he said, appealing to White. "You need to talk to Systrom."

Over the next few weeks, the more White discussed Instagram's future with Systrom, the more she realized that she wanted to work with him. She'd been early at Google, early at Facebook, and here was a way to be early at Instagram without leaving the company.

Some of her fellow Facebook executives advised her against the move. They said this role was simply too small at a time when her career held such promise. And friends of Sandberg, or "FOSes," as they were known internally, had a reputation for not shining as brightly once they were out of her dominion—at least according to the mostly male staff. White ignored the pushback. *We're about to piss away a billion dollars and a fabulous team because no one in the larger company really understands what we just bought,* she thought.

After the turmoil, Systrom got his CEO title back, since Facebook wanted him to have authority to sign off on independent decisions.

· · · · · · **·** · · · · · **·**

Systrom was relieved to have someone who could help him understand how to build a company within Facebook. He and White met for several hours each week, trying to figure out how to articulate ways Instagram was different, what they needed help with and what they didn't. They surveyed Facebook employee phones and found out that only about 10 percent of them even used Instagram, similar to the rate of the broader U.S. population. The first step, then, would have to be education.

White hired a designer to come into Instagram's building and mount all the photography books, old cameras, and bottles of bourbon on shelves, to make the space a bit more crafty and thoughtfully displayed. (Friends and business partners always gave Systrom bourbon as a gift, as a tribute to the early app's name.) The design provided contrast to Facebook, where the "journey is only 1 percent finished" motto was reflected physically in open ceilings, exposed pipes, and unvarnished wooden surfaces. Once a week, Instagrammers would roll up their garage door and invite passing Facebookers in for coffee, in an attempt to make friends.

(While there was free coffee everywhere on campus, Instagrammers could offer *good coffee*, the kind that came from the pour-over kits and espresso machines they had learned to prefer.)

Together with Krieger, Systrom and White came up with a mission statement that the *Wall Street Journal* would later call lofty and hokey: "To capture and share the world's moments."

White recruited to Instagram's job slots new employees from the Facebook side, who brought with them a devotion to analytics. But the same hacker mentality that was rewarded at Facebook caused clashes among the expanding Instagram team. Former Facebookers would give obvious ideas to boost activity, like adding a re-gram button, and then original Instagrammers would reject them, saying, "That's not how we do it here." Instagrammers would explain the charm of InstaMeets or discuss a plan for highlighting the Albuquerque International Balloon Fiesta on the @instagram account, and some former Facebookers would roll their eyes.

But how *did* Instagram do Instagram? The original employees of Instagram worked together to figure out the best way to explain to their new Facebook coworkers what their culture was like. They brainstormed and researched, at one point even asking members of a focus group to draw a picture of what Instagram would look like if it were a human. (They mostly drew male faces with sideswept bangs and dark eyes; the illustrations looked eerily like Joshua Riedel, the first employee, who was still there.)

Ultimately the team came up with three Instagram values, all of which included not-so-subtle notes of culture clash with Facebook.

The biggest was "community first," meaning all their decisions should be centered around preserving a good feeling when using Instagram, not necessarily a more fast-growing business. Too many notifications would violate that principle.

Then there was "simplicity matters," meaning that before any new products could roll out, engineers had to think about whether they were solving a specific user problem, and whether making a change was even necessary, or might overcomplicate the app. It was the opposite of

Facebook's "move fast and break things," where building for growth was valued over usefulness or trust.

There was also "inspire creativity," which meant Instagram was going to try to frame the app as an artistic outlet, training its own users and highlighting the best of them through an editorial strategy, focusing on content that was genuine and meaningful. This was a rejection of the self-promotional fakery that was already starting to define some of Instagram's popular accounts. It was also a very different strategy than Facebook's algorithmic personalization approach. "We don't have a voice," Chris Cox, the head of the news feed, would tell employees. "We give people a voice."

· · · · · ● · · · · · ·

The community team at Instagram—the team focused on writing blog posts about interesting accounts and supporting user events—violated another central Facebook tenet, which was that Facebook only concentrated on things that *scaled*. They didn't have outreach to their power users because a group, no matter its influence, didn't matter strategically as much as the whole. What's the return on investment for supporting one person, or several dozen people, when you could instead deploy your resources in a way that affects hundreds of millions, or even billions?

Instagram considered its community team to be the soul of the place, doing work that helped set the tone for the rest of the millions of users. Whatever they highlighted on the @instagram account would be either followed or mimicked by others. They also kept tabs on the ways the product was used differently in various countries, alerting Instagram's product managers about the requests, struggles, and opportunities they saw. They still rotated names on the suggested user list to highlight potential new interests for new members to follow, and ran their blog on Tumblr.

The work highlighted their ideal version of Instagram: people using the app to showcase the way they were grinding their own matcha in Kyoto, or hiking Mount Kilimanjaro, or designing their own canoes in coastal Oregon. The editorial strategy highlighted people approaching the product in new ways that would inspire Instagram users. Instagram explicitly encouraged this with contests like a weekend hashtag project, for which they asked users to post images of a #jumpstagram, a mid-air jump—or a #lowdownground, images shot from a perspective on the floor. Thousands of entries were submitted each week for the chance to appear on the @instagram account.

Instagram users, feeling like they had a relationship with the brand, were still hosting their own InstaMeets in different parts of the world to make new friends and talk about photography. Some were even crafting their own physical replicas of the Instagram logo, out of arranged flowers, hand-knit blankets, or decorated cakes. The value of user obsession was difficult to objectively quantify, or to tie back to the team's editorial efforts.

Zollman and White would get into fights about the return on investment for user outreach, to the point that Zollman quit before her one-year bonus time, sensing that her contributions were no longer valued. And she had other reasons too: the shuttle commute, the fact that she couldn't bring her dog to the office, that employees were no longer hanging out like they used to. Mostly, she hated Facebook's metrics-based employee review process. How could she show she was driving growth if she was just in charge of inspiring people?

Before she quit, Systrom listened to Zollman's concerns, but didn't intervene. He knew that if Instagram wanted to truly be influential within Facebook, if they wanted to prove they deserved that generous acquisition offer and all the resources, including for the community team, they needed to do something that Facebook would value. Instagram needed to either be quashing competitors or making money. He figured the money part would come pretty naturally on the app if they could get it right, since a visual medium was alluring and aspirational, which made it perfect for selling things and building brands—as long as it didn't look like traditional advertising.

Systrom went to Zuckerberg with ideas for building revenue but was quickly shot down.

"Don't worry about that right now," Zuckerberg said. "Just keep going. All you've got to do is keep growing."

Then Systrom appealed to Bosworth, the ads vice president he'd picked a fight with the prior year. "No, man," said Boz, who respected Systrom's ambition, and he started to become fond of him. "We don't need you right now. You've got to grow." Facebook's mobile advertising was starting to show promise, and so Systrom needed to follow Zuckerberg's thesis, that moneymaking should come only after the network had staying power.

Despite the discouragement, Systrom spent hours brainstorming with White and Amy Cole, the early business employee, about what a strategy might look like, whether in commerce, advertising, or something else entirely. Until that happened, he and Krieger decided, it was time for Instagram to execute on one of Zuckerberg's other priorities. It was time to address a competitive threat.

Systrom thought about his counterparts at other acquired companies. Tony Hsieh, the CEO of the online shoe business Zappos, hadn't gotten to remain in Jeff Bezos's orbit after Zappos was acquired by Amazon in 2009. YouTube's founders weren't even relevant to YouTube anymore—they'd left the company after the 2006 Google acquisition.

He had no intention of being forgotten like that.

DOMINATION

"We're looking to have a level of impact on the world that is unmatched by any other company, and in order to do that we can't sit around and act like we've made it. We need to constantly remind ourselves that we haven't won and that we need to keep making bold moves and keep fighting or we risk peaking and fading away."

—MARK ZUCKERBERG, QUOTED IN
THE FACEBOOK EMPLOYEE HANDBOOK

Perhaps Zuckerberg was comfortable giving Instagram some level of independence because he saw so much of himself in its founder. He and Systrom were, on paper, quite similar.

Both were raised in comfortable suburban homes by loving, married parents and had siblings they were close with. Both attended elite East Coast boarding schools and top private universities, where they became fascinated not just with engineering, but with history—in Zuckerberg's case, the history of Greek and Roman empires; in Systrom's, art history and the Renaissance. They were basically the same age: Systrom was older by five months, but Zuckerberg seemed to have more wisdom after running his company for longer.

Still, their relationship was businesslike, with Systrom trying to make Instagram important to the overall company without losing his hold on its future. The men would have a strategy dinner every month or so at Zuckerberg's home, but in practice Zuckerberg's home was another office. Following the 2010 movie *The Social Network*, he'd had to invest more deeply in personal security, as he was unable to go anywhere in public or fly commercial without being recognized immediately. In 2013, he spent $30 million to purchase the homes surrounding his own in Palo Alto, to afford himself more privacy.

Zuckerberg's house wasn't just for business meetings. He did host social gatherings there, just not ones Systrom was invited to. There was a crew of Facebook employees—like ads leader Andrew Bosworth and news feed boss Chris Cox—who would be invited over for weekend barbecues with their wives. These friends had been through the turbulence of Facebook's early days, back when Facebook provided a $600-a-month rent stipend to anyone living within a mile radius of the office, then in downtown Palo Alto. Early employees ended up building their lives in the same neighborhood, working hard, and then socializing together and posting about it on Facebook.

These barbecue guests were the same people present at Systrom's Monday leadership meetings, where he was struggling to have a voice. They comprised a clique Systrom would not be part of, just like everyone would revere but not quite understand the power of his relationship with Krieger.

At one point, Zuckerberg took Systrom on a ski trip, attempting to bond with him. But the outing only served to display the differences between the men and their egos.

Systrom was competitive, but it was always very important to him to do things *the best way*. He would pick his wine from the highest-rated year, he would try to absorb knowledge from the most talented people, and he would read stacks of books about whatever new skill he hoped to master. He'd soon have a personal stylist, a personal trainer, and a management coach. He would drink coffee made from Blue Bottle beans only at their peak point—four days after roasting. "I have a special machine for it and a scale that reads out the extraction by the second, so you get a graph," he later told the online journal of the fashion brand *MR PORTER*.

When he was a child, his father brought home a bat, ball, and mitt so they could practice baseball in the backyard. Systrom asked if he could go to the library first, so he could check out books about pitching technique before playing.

Zuckerberg, on the other hand, was set on doing things *better than anyone else*. He loved board games, especially strategy games like Risk. In the early days of Facebook, he would occasionally play in the office, tweaking his technique so his opponents could never predict his next move. He once lost to a friend's teenage daughter while playing Scrabble on a corporate jet, and was so frustrated he built a computer program to find him all the word options for his letters.

When Google launched a competing social network in 2011, Zuckerberg rallied Facebook employees into action by quoting ancient Roman senator Cato the Elder: *"Carthago delenda est!"* Or "Carthage must be destroyed!" Then, as he often did at Facebook when there was cause for alarm, he would institute a "lockdown," requiring people to work longer hours, and spin up "war rooms"—conference rooms dedicated to winning competitive fights. There were war rooms for everything.

On the ski trip, Systrom was checking an app called Ski Tracks that showed him the length of his run, the altitude, the angle of the slope, and more. He'd downloaded it to improve the quality of his performance.

"What's that?" Zuckerberg asked. "Does it show you top speed?"

It did show top speed.

"I'll beat you down the next hill!" Zuckerberg declared, making Systrom instantly uncomfortable.

Systrom preferred backcountry skiing, challenging himself with unpredictable terrain, but Zuckerberg, ever since his younger days, loved racing downhill. And even on the mountain, he was the boss.

Companies become a reflection of their founders. Systrom had created a place on the internet where the most interesting people who were the best at what they did could be followed by others, praised, and emulated. He chose to grow that community with an editorial strategy that drew attention to top talents. The product was simply a venue for what its users were doing, and Systrom didn't want to make big adjustments and risk ruining it, unless the change allowed the app to remain a high-end product experience.

Zuckerberg had created the largest network of humans ever. He chose to grow that community by tweaking the product constantly to pursue a greater and greater share of the time people spent on the internet, meanwhile looking at what his competitors were doing and coming up with strategies to undermine them.

Systrom had never met anyone as tactical as Zuckerberg. He wanted to learn Zuckerberg's ways, but also to assert that he was a CEO—one of the good ones—in his own right, in a way that didn't have to be so aggressive. His next move would appeal to Zuckerberg, helping him see Instagram as a useful partner. But Instagram on its own was not enough to satisfy Zuckerberg's zeal for industry domination.

- - - - - - - - - - - - -

The Instagram acquisition had a tremendous ripple effect on the rest of the industry. Other social media apps were suddenly getting investor attention, with the idea that they too might one day be acquired by a Facebook or Twitter for a rich sum.

Facebook started out with text; Instagram started out with photos. The next generation of social apps was all about video. Users had long been asking Instagram to launch video, to the point that venture capitalists funded a handful of startups to beat them to the punch, including Viddy, Socialcam, and Klip. YouTube and Facebook had video but weren't built naturally for mobile phones. Still, Instagram didn't make a move until it had to.

Twitter, after losing out on the Instagram acquisition, bought the next up-and-coming app that Jack Dorsey suggested: Vine, which people would use to produce and share six-second videos that looped over and over. Twitter purchased Vine several months before the app's January 2013 launch.

Most people didn't have something they wanted to film for just six seconds. But Vine's constraint would inspire new types of activity, just like Instagram's square requirement or Twitter's 140-character limit. Creative people figured out how to use Vine to showcase their perfect comedic timing or shocking tricks. They flocked to the new app, amassing audiences that made them small stars for their skits. Some of them, like King Bach, Lele Pons, Nash Grier, and Brittany Furlan, were drawing millions of followers. Twitter had no idea what to do with the product, just like Facebook wasn't sure what to do with Instagram.

Systrom, always careful with quality, had told people he wasn't interested in video yet because phone connections were too slow for a good experience. Vine proved that wasn't a problem anymore.

"We don't want Vine to be the Instagram of video," Systrom started saying. "We want Instagram to be the Instagram of video." Systrom and Krieger gave their engineers a six-week window to build and ship a way to post 15-second videos in the Instagram feed. There was no Facebook-style optimization baked into that number of seconds. It was "an artistic choice," Systrom would say.

Having a singular mission rallied Instagram's troops out of their post-acquisition malaise, the way wartime causes citizens of a country to become more patriotic. Krieger was especially grateful for the opportunity to build something, as opposed to spending all his time fixing infrastructure to address the app's rapid growth. As they worked on the video project, he also taught himself how to be a better Android engineer to help the team meet its deadline.

Android phones were notoriously more difficult to build for, since they came in different sizes by different manufacturers. Krieger spent the night before launch working through the bugs with the Android leads, testing the app on various versions of the phones they'd ordered on eBay past 3 a.m. The assembled group decided to sleep at the offices. One engineer pulled couch cushions together in an empty conference room. By 5:30 a.m., Krieger could be found barefoot in the office bathroom, brushing his teeth.

On launch day, Facebook corralled the press into a room that had been completely redesigned to look like a coffee shop, with newspapers strewn about on tables, in a nod to Instagram's ubiquitous latte photos. Zuckerberg gave some opening remarks, then handed the floor to Systrom. The gesture was quite symbolic: here was Zuckerberg, deciding *not* to be the main speaker at a Facebook product launch event, and letting it all be Instagram-branded. The small team had earned some respect.

Afterward, Zuckerberg, Systrom, and all the others went back to the Instagram office and watched a ticker count up the number of videos posted. It was the first (and last) time anyone remembered Zuckerberg coming to the Instagram office. They all cheered when the count reached 1 million.

Krieger, running on little sleep, scrolled through his feed and saw a post that made him well up with tears. A Japanese friend he'd followed since the very early days of the app, who had a very adorable dog, had posted a video. It was the first time Krieger had ever heard his friend's voice.

Instagram had done something that was important not just for relationships on the app, but also for Facebook Inc. And so they finally had a reason to celebrate—the way Systrom wanted to celebrate. The team went on a retreat to Sonoma wine country, where they stayed at the Solage resort, rode in hot-air balloons, ate food cooked by a celebrity chef, and took joy rides in rented Mercedes convertibles.

• • • • • • • • • • • •

Systrom and Krieger expected video would become another common type of post for regular people, like the Japanese man with the dog. But as was evident from Vine, most people didn't have a reason to post brief videos unless they had something really specific to show, like cake decorating, fitness routines, or short-form skit comedy.

So the top people who paid attention to Instagram video were the same people who grew their followings on Vine. Many of them were helping each other, cowriting and filming skits in Los Angeles, hanging out in Darwyn Metzger's office on Melrose and Gardner in West Hollywood. Metzger's company Phantom would give them space to collaborate, while helping them negotiate deals to make Vines for brands. Viners like

Furlan, Marlo Meekins, and Jérôme Jarre balked at the idea of working with businesses, thinking their audiences would hate them for selling out. But eventually the price was right, and the small stars started to become dependent on the income, with the biggest names on Vine making thousands of dollars per post.

Metzger knew it was unsustainable, partly because he didn't trust Twitter's leadership. The day Instagram launched video, his fears were realized. *Anything competitive with the power of Facebook behind it means Vine is seriously doomed,* he thought. So he told his crew, "From now on, you have to take one-third of your day and start migrating your audience somewhere else. I don't care if it's Instagram or if it's YouTube or Snapchat, but you need an alternative to Vine."

While it was hard news to swallow, they took his advice. Several former Viners, including Furlan, Pons, and Amanda Cerny, started transitioning their efforts to Instagram, where they eventually drew followings in the millions.

• • • • • • • • • • • •

With the video strategy, Systrom had bet correctly that crushing competition was the best way to win over his new Facebook overlords. But he had underestimated Zuckerberg's paranoia. Unbeknownst to him, Zuckerberg was interested in finding other Instagrams to buy. It turned out that his big purchase was just part of a larger strategy to own multiple apps and place more competitive bets, hedging against Facebook's inevitable fade, which Zuckerberg thought could come at any time.

While welcoming Systrom into his company in 2012, Zuckerberg was emailing another young man, who was building a different app that appeared to be a breakout success. He also had elite schooling and a charmed upbringing, at least financially. His competitive philosophy? That *everyone else was doing it wrong.*

Evan Spiegel's Snapchat app started out as a Stanford party tool in 2011, as a rejection of the world Facebook and especially Instagram had created. When everything people posted was polished up for public consumption with likes and comments, where was the fun? Where was the place to put all the debaucherous things twenty-somethings were doing, that didn't need to end up on their permanent social media record, staining their job prospects? As the Kappa Sigma fraternity member in charge of hyping parties, he saw an opportunity.

With help from fraternity brothers Bobby Murphy and Reggie Brown, he came up with an app that was all about sending a photo that would vanish after a few seconds. The first version was called Picaboo. "It's the fastest way to share photos that disappear," Spiegel wrote in a pitch email to fraternity website BroBible, with the subject line "Ridiculous iPhone App," calling himself a "certified bro." He explained that you take a picture and set a timer for up to ten seconds; once your friend opens the message, it lasts for that time, and then it's gone. "Fun shit," he added.

Spiegel, tall and thin with cropped sandy hair, straight eyebrows, and a dimpled chin, was as irreverent as Systrom was careful. Spiegel had grown up an introvert, finding it difficult to trust people, preferring the comfort of luxury cars. He was the son of a powerful corporate lawyer who had just defended Transocean Ltd., the company that owned the oil rig responsible for the 2010 BP spill in the Gulf of Mexico.

Besides having a tendency for profanity, Spiegel was apt to pick fights and hold grudges. Brown later sued, claiming Snapchat ousted him from the company and didn't give him credit for being a cofounder. Snapchat settled.

But irreverence was appealing in Snapchat's product. Spiegel hated having to think about other people's opinions on his life or his decisions, and he wasn't alone. Online personal brands were becoming more important to cultivate in society, and therefore more anxiety-inducing. Picaboo made few waves. But when the founders rebranded as Snapchat and added video, plus the ability to draw and write on photo and video messages with digital markers, they made something less stressful and more fun, and much more appealing to young people.

"People are living with this massive burden of managing a digital version of themselves," Spiegel told *Forbes* writer J. J. Colao. "It's taken all of the fun out of communicating."

At first, Snapchat was described by the media as a sexting app. If you weren't sending nudes, why else would you need your photos to disappear? But that characterization misunderstood how teens were using technology.

Instagram's reality-warping filters and curated, crafted feel had a downside: pressure. For Instagram, teens were filling their camera rolls with dozens of different angles of the same shot, finding the perfect one, then editing away their imperfections before posting. They were going out of their way to do things that were cool and visually interesting. And they would often delete pictures if they didn't get 11 likes. That was the number of likes that would turn a list of names below an Instagram post into a number—a space-conserving design that had turned into a popularity tipping point for young people.

Snapchat was a different world. Young people were sending each other random selfies and unedited videos. The app was confusing for adults because it wasn't for sitting and scrolling through content—it opened directly to a camera mode, which was for capturing and sending whatever was happening right in that moment. Snapchatting was like texting, or having an asynchronous video chat conversation. And it was fun.

"The main reason that people use Snapchat is that the content is so much better," Spiegel said to *Forbes*. "It's funny to see your friend when they just woke up in the morning."

Older people weren't supposed to get it. By November 2012, Snapchat had millions of users, most of them between 13 and 24 years old, snapping 30 million times a day.

• • • • • • • • • • • •

Spiegel's app could have faded from the market—or he could have been kicked out of his father's house after dropping out of school. But after Facebook acquired Instagram, everything changed. Cash from investors was suddenly easy to get. So was respect, and attention from acquirers.

That November, while still navigating the Instagram integration, Zuckerberg was back on the hunt. He sent an email: "Hey Evan, I'm a big fan of what you're doing with Snapchat. I'd love to meet you and hear your vision and how you're thinking about it sometime. If you're up for it, let me know and we can take a walk around Facebook HQ one afternoon."

Snapchat's appeal with teens was crucial. Teens, about to leave high school and enter the wider world, were quickly building networks that would serve as infrastructure for the rest of their lives. At that age, they were building new habits and amassing spending power without oversight from their parents, developing affinities for brands they'd have loyalty to for years. Facebook might have started with college students, but Zuckerberg knew it needed power among this younger cohort.

His email to Spiegel was the same kind of suggestive-without-saying-anything outreach that Systrom had received from tech giants when the Instagram app first started blowing up. Spiegel subtly played hard to get.

"Thanks :) would be happy to meet—I'll let you know when I make it up to the Bay Area," he replied.

Zuckerberg responded, saying he would just happen to be in Los Angeles soon. He had to meet Frank Gehry, the architect who was going to design another building on the Facebook campus. Could they meet near the beach? Spiegel agreed, and he and cofounder Murphy met Zuckerberg in a private apartment that Facebook rented for the occasion.

Once they were together in person, Zuckerberg abandoned the flattery and went straight to threats. He spent the meeting insinuating that Snapchat would be crushed by Facebook unless they found a way to work together. He was about to launch Poke, an app that would allow people to send disappearing photos, just like they did on Snapchat. He was not afraid to completely copy their product, putting all of the power of Facebook behind making it a success.

It was flattering that Zuckerberg, the king of the Internet, considered Snapchat a threat. Spiegel was onto something.

• • • • • • • • • • • •

The day Poke launched in December 2012, it at first showed the power of Facebook's endorsement. Suddenly in front of millions of people at once, it became the top free app in the iOS app store.

And then, starting the next day, it declined and declined in the rankings. Zuckerberg's threat turned out to be empty. Even worse for him, many of the people who downloaded Poke, who hadn't known of Snapchat before, became aware through the process that there was another app doing the same thing better. Snapchat's downloads climbed.

Facebook had copied Snapchat's functionality but they had failed to copy the app's cool factor. They were facing the same problem they'd had when trying to build a camera app that copied Instagram. The social networking giant could harness the attention of millions, but the quality and feel of the product had to do the rest of the job.

Luckily, Facebook had another tool in its arsenal: money, and Zuckerberg's unilateral power to make decisions with that money. He offered to acquire Snapchat for more than $3 billion. It was even more shocking than the Instagram price for about the same number of users, and was also heavily weighted with Facebook stock, which was climbing back to its $38 IPO price.

Just as shocking, Spiegel declined. The 23-year-old CEO sensed weakness, and therefore opportunity. More importantly, he and cofounder Bobby Murphy had no interest in having Zuckerberg as their boss.

In June 2013, Spiegel raised $80 million from venture capitalists instead, valuing the company at more than $800 million, after less than two years and without any revenue, and with just 17 employees.

Zuckerberg, frustrated that so far he could neither build nor buy what Snapchat had, resolved to get a lot better at understanding teens, why they had fled Facebook, and how he could recruit them back.

• • • • • • • • • • • •

The ordeal confirmed Spiegel's suspicions that Facebook was for the olds, and would one day fade into being the next Yahoo! or AOL. He wanted to be nothing like them. He banned employees from using words like "share" and "post" that reminded him of Facebook, since Snapchat was about being more personal, and preferred using a term like "send" instead.

He was determined to keep releasing ideas that Zuckerberg would never think of. What if Snapchat had an option to "send to all," where content would still disappear, perhaps after 24 hours? Spiegel had come up with the idea while still in college, calling it "24 Hour Photo," after the stores that take a day to develop film. He was brainstorming with Stanford friend Nick Allen about allowing multiple photos, so people could create a flip-book for their days. On Instagram, you just posted the *best* picture or video from the party. But what about the photos and videos from getting ready, heading to the event, encountering friends there, and then being too hungover to go to class the next day?

The Snapchat team had graduated out of Spiegel's father's mansion and into a tiny blue house on the Venice Beach boardwalk in Los Angeles, where interesting things were happening all the time. Stoners were skateboarding by, hippies were making art with cans of spray paint, beautiful beachgoers were suntanning. The backdrop made it easy to imagine that one of the most pressing problems in media was not having enough ways to show everyone what was going on.

Allen, who had joined the company after graduating that year in 2013, explained the specifics of the vision to the engineers: the product, called Stories, would be organized chronologically, with the oldest post appearing first, unlike on Twitter and Instagram, which always showed the most recent post first. Each addition to the Stories queue would expire after 24 hours. If users looked in time, they would see a list of the names of every single person who had checked out their update.

Snapchatters would not "post to" their Stories; they would "add to" them. But now, with a broadcast tool, one that lowered the bar for what was good enough to capture on social media, Snapchat created a habit for the same young people they hoped to free from pressure.

• • • • • • • • • • • •

Meanwhile, Systrom had no idea Zuckerberg was talking to Snapchat, much less threatening them and trying to buy them. As Zuckerberg started emphasizing teens on the platform, Systrom felt ahead of the game. Teens weren't on Facebook, because their parents were there. Parents weren't on Instagram yet, and thanks to Instagram's new emphasis on data, they knew that the app's demographic breakdowns showed young people were Instagram-obsessed.

After successfully launching video, Instagram was feeling independent within the Facebook ecosphere. Or were they just being ignored? Systrom was good at spinning the situation to sound nice. *I'm the CEO of Instagram, and we're basically a company still, with Zuckerberg as our board member,* he'd say.

But Instagram wasn't much of a company if it relied wholly on Facebook's advertising revenue. Systrom was still a CEO who didn't make money. A few months after Zuckerberg had told Systrom to hold off on a business model, Instagram had proved itself more, with a user base well past 100 million. So in the middle of 2013, Facebook was finally willing to let the team experiment with ads.

Systrom and his business team decided that if advertising was going to work on Instagram, promotions needed to look like Instagram posts and be visually pleasing, casually artsy, without trying too hard to sell; there could be no writing or price tags on the image itself. It was important, as Systrom had said the prior year, that any post from a brand "comes across as honest and genuine." Instagram modeled the look off *Vogue* magazine's: high-end brand advertising showcasing products in a subtle manner, as just one element of the lives of beautiful, happy people.

That September, Emily White was featured in the *Wall Street Journal*, with the headline "Instagram Pictures Itself Making Money." The writer, Evelyn Rusli, compared White's role at Instagram to Sheryl Sandberg's at Facebook. Rusli reported that White was spending her weeks meeting big-name advertisers like Coca-Cola and Ford Motor Co. and "wanted to avoid repeating some of Facebook's earlier missteps with advertisers," a line that ruffled feathers internally.

But Facebook ads and Instagram's ad plan stood in sharp contrast. Facebook sold ads through an online system that anyone with a credit card could participate in. Even the top brands, some of which had help from Facebook salespeople, still had to buy ads through this open system.

It was built this way so that anyone could pick and choose what kind of audience they wanted to see the ad, with the more specific or in-demand audiences costing more. Those audience choices were automatically matched up with users who fit the profile. Facebook employees weren't reviewing or even looking at ads before they went up, except in rare cases.

Instagram, on the other hand, was trying to build a premium experience, brainstorming directly with advertisers about their ideas and manually placing their ads. They knew that this system couldn't work forever, but Systrom and Krieger always urged people to do the simplest thing first, the way they had when they first built the app. Working manually on a small version of the product made more sense than spending precious engineering resources and navigating politics with Facebook's ads sales team, for a system that might not ultimately work.

Using a strategy similar to that he'd employed when he founded the company—picking launch partners like Burberry and Lexus who would *get it*—Systrom personally approved every ad. Especially since now Instagram's brand was too precious to risk letting anyone and everyone advertise however they'd like.

· · · · · · · · · · · ·

Instagram ran its very first ad on November 1, 2013. Michael Kors, one of the premium brands the team had lined up, was allowed to post a photo on the @michaelkors account and then pay to distribute it to people who weren't already following. The image looked like it was straight out of a glossy lifestyle shoot in a fashion magazine: a gold watch with diamond trim, placed on a table surrounded by a gold-rimmed teacup and colorful French macarons. A green macaron had a bite missing, giving a sense that it wasn't just a prop. "5:15 PM: Pampered in Paris #MKTimeless," the caption said.

Only one brand per day, Systrom had decided—that felt right. It was nonnegotiable: if Louis Vuitton called wanting the twentieth of the month, they would decline if Ben & Jerry's already had the slot. All the names of the early advertisers were mapped out in red marker on a whiteboard calendar. An employee would print the potential ads out; then Systrom would go through them, one by one, deciding what was good enough and what wasn't. If an ad wasn't good enough, he would protest.

At one point Systrom was concerned that the food in one of the branded posts looked unappetizing, especially the French fries, which appeared soggy. "I don't want to run it like this," he told Jim Squires, his new ads lead, who had come over from Facebook.

"Well, we have urgency to run this for the client," Squires said.

"No problem," Systrom replied. "I'm on a flight this morning. I can fix the white balance and sharpen it up." After he made the potatoes look crispier, he sent the photo back to Squires over Facebook Messenger, and then the ad ran.

Systrom's focus on the quality of the photos rather than on the readiness of Instagram's technology caused problems. On that first day, representatives from Michael Kors called to complain that the hands on the watch actually read 5:10, not 5:15. They didn't know how to edit their caption. The Instagram team confessed that so far, there was no way for any user to edit their captions, and no way for the company to override and do it for them. The error would have to stand. But the press writing up the news of Instagram's advertising launch didn't seem to notice.

· · · · · · • · · · · ·

In order to launch the advertising business, Instagram had to dodge an uncomfortable reality: advertising agencies hated Facebook. Teddy Underwood, an early Facebook employee who had just transitioned to Instagram to promote its new advertising products, thought the only way to sell them was to make a case that Instagram was the anti-Facebook. He set up meetings with the largest ad agencies, armed with a polished PowerPoint presentation about the value of inspiration. He told them Instagram was completely independently run, didn't plug into Facebook's ad system at all, and had a plan to build better relationships and effective ads, suited to their audience and aesthetic.

There was something awkward about his role, though. Instagram's Emily White was only sort of his boss. Carolyn Everson, the new head of sales at Facebook, was the one in charge of advertising strategy. A lot of people on the sales and marketing side of Instagram had double bosses like this. The independence Zuckerberg had promised Instagram was holding for the product and engineering side, but the sales and operations side, run by Sheryl Sandberg, was starting to assert a deeper level of control.

One day, Underwood went into a conference room to report progress to Everson via video call. His pitch to make advertisers think Instagram ads were worth more than Facebook's had worked—and had resulted in a major deal with one of the four big ad agencies.

"Omnicom has committed $40 million in Instagram advertising next year," he reported, "and I think one of the other big agencies is willing to commit soon."

He didn't get the reaction he expected. It turned out Everson had been looking for a way to get ad agencies back on Facebook's side and was looking to use Underwood's success to help the company overall. "Instagram clearly is the shiny object right now that agencies can't have and really want," Everson said via video conference from New York. She was surprised that Instagram was able to get such a large commitment so quickly and wanted to make use of it. "We have more leverage than we thought we did."

She asked Underwood to go back to the ad agency and say that they would only get the $40 million on Instagram if they committed to $100 million on Facebook as part of the deal. Underwood refused, saying he valued his relationships and had promised a new kind of ad—not more of Facebook. Everson said the Facebook team would handle it from there. In fact, she insisted that future Instagram ads not be sold by a separate team at all. Underwood, realizing the Instagram job wasn't the return to startup life he'd expected, didn't last much longer in the role. Everson didn't get exactly what she wanted either. When the Omnicom deal was announced in 2014, it was just for Instagram ads. Everson would later deny she ever asked for more.

●　●　●　●　●　●　●　●　●　●　●　●

Facebook, determined not to get complacent about its dominance, even when surrounded by underdogs, was always looking for a way to push a little further. The company had Instagram reduce visibility for the #vine hashtag on Instagram, and discouraged prominent users from displaying their Snapchat usernames. And even when they couldn't control the competition like they could Instagram, they could still study it—in detail.

Facebook in 2013 acquired a tool called Onavo. The acquisition generated little buzz, as it wasn't a flashy consumer product. It was a wonky-sounding thing called a virtual private network, or VPN, which was made by Israeli engineers to allow people to be able to browse the Internet free from government spying on their activity, and from having to go through firewalls.

For Facebook, the acquisition was crucial. While people were escaping the watchful eye of their governments, they were unwittingly giving Facebook competitive intelligence. Once Facebook purchased the VPN company, they could look at all the traffic flowing through the service and extrapolate data from it. They knew not only the names of the apps people were playing with, but also how long they spent using them, and the names of the app screens they spent time on—and so, for example, could know if Snapchat Stories was taking off versus some other Snapchat feature. It helped them see which competitors were on the rise before the press did.

The data was easily accessible to Facebook employees, and was funneled into regular reports for executives and the growth team so that everyone could keep tabs on the competition. It was the first thing Emily White checked when the Wall Street Journal broke the news, many months after Zuckerberg's meetings, that Facebook had tried to buy Snapchat for $3 billion. And it was the first thing she thought about when she got an aggressive message from a recruiter on her cell phone.

The recruiter told White he had a once-in-a-lifetime chief operating officer job for her and that if she didn't call him right back, he would never call her again.

"Listen," she said when they connected, "I would love your help at some point. But in like five years, not now."

When she hung up, she thought about what he'd said. The recruiter had mentioned it was a fast-growing consumer-facing startup that wasn't in Northern California. White realized she knew exactly which company he was talking about, and started to let herself get a little excited.

She had spent almost her entire career working under Sheryl Sandberg, at Google and then at Facebook. Half her time at Instagram was wrapped up in navigating internal politics, and she wondered what she'd be capable of outside Sandberg's purview. But she didn't want to leave for a competitor.

The Onavo data showed that the app usage for Snapchat and Instagram was not competitive but was positively correlated: if someone used Instagram, they were likely to use Snapchat too. White reasoned that perhaps Spiegel's startup was filling a void in social media, creating a place where people could be casual, as a complement to what they could accomplish on Instagram.

She talked to her husband. "People who don't take risks work for people who do," he told her. She called the recruiter back and said she was interested.

• • • • • • • • • • • •

The data wasn't telling White the whole story about competition. Snapchat, in fact, was growing into the first serious threat Instagram had faced since its early days. Snapchat had just launched Stories, offering broadcasting to a larger feed than just direct messages. And Instagram was about to launch direct messaging, their first attempt to try a tool that was about sending posts to one person versus broadcasting to an entire feed.

When White resigned for the Snapchat COO job, it rattled Systrom's confidence. He had spent so many days brainstorming with her, traveling with her, planning the business model with her guidance. Accepting her into the executive role was akin to embracing and trusting Facebook. Now he was, in a sense, paralyzed with doubt about his own decision-making, specifically around who he had decided to trust. Most of the people White had hired for Instagram were former Facebookers. After her departure, Systrom stopped holding question-and-answer meetings with staff for a while. For a couple months, he started showing up to work later, and paused some of his hiring plans.

Zuckerberg too was managing his own concerns. Far removed from White's departure, he worried, as always, about Facebook's continued pursuit of domination, fighting off an inevitable irrelevance. Facebook was about to turn ten years old as a company, and almost half the world's internet-connected population was using the product. The proportion was larger if you didn't count China, where Facebook was blocked by the government. So, assuming they kept going and added a larger proportion of the world, what then? If Facebook had more Snapchatesque rejections, if they couldn't buy more Instagrams, how else could they grow?

First, he tried to get his own employees to build more interesting Facebook competitors, inside Facebook. He couldn't count only on the Onavo intelligence to warn him early about up-and-coming products, or assume that he would be able to buy them. Facebook also needed to try to create the next Snapchat, or the next Vine, themselves. That December 2013, the company hosted a three-day hackathon—an event just for coding new app ideas—to kick off an entirely new initiative at the company called Creative Labs, which would be their internal startup accelerator. About 40 ideas emerged from the event, none of which would be much more successful than Poke, which eventually died.

Second, Zuckerberg launched an initiative to add more people to the internet, all of whom could be potential future Facebook users. He started a division of Facebook called Internet.org, which sounded like it was a nonprofit; it would be charged with figuring out how to bring connectivity to remote areas of the world, using drones, lasers, and whatever else its team could come up with.

And third, Zuckerberg realized he had another secret weapon: Systrom himself. Just as Instagram advertising had allowed Facebook an opportunity to repair relationships, Instagram's apparent independence within Facebook could be a selling point to founders on the fence about joining. Systrom led a life that was enviable to other founders Facebook wanted to bring onboard. To anyone in a similar position to Systrom in 2012—with a popular product and a business model that was either uncertain or nonexistent—Zuckerberg could offer a way to continue running things and keep their CEO title, with no financial risk and all of the network and infrastructure Facebook could provide.

After the failure with Snapchat, Zuckerberg asked Systrom to help acquire the app he wanted to pursue next: WhatsApp, the messaging app that had 450 million monthly users all over the world. According to Onavo data, the app thrived especially in countries where Facebook wasn't as dominant.

Systrom dutifully helped Zuckerberg sell the vision. In early 2014 he had a sushi dinner with Jan Koum, WhatsApp's CEO, at Nihon Whisky Lounge in San Francisco. Systrom helped reassure him that Facebook was a good partner, unlikely to ruin what made WhatsApp special.

Koum was notoriously untrusting, after growing up under surveillance by the USSR in Ukraine. He built an app that was end-to-end encrypted, so the records of what people were saying to each other weren't readable by anyone—not the police, and not even his company. He promised his users "no ads, no games, no gimmicks," just a simple tool they could pay $1 a year to use. It would be a stretch to join Facebook, where surveillance of users powered the advertising engine.

Systrom said enough to help convince Koum that Facebook's promise of independence was real. He and cofounder Brian Acton would be able to preserve their values at the social networking company, despite its advertising business model.

The money was perhaps even more convincing to Koum than Systrom was. When the deal was announced, everyone at Instagram was shocked all over again. The price was a stunning $19 billion. Plus, Koum got a seat on Facebook's board, and WhatsApp got to stay in its own offices in a nearby town called Mountain View, with about fifty employees who were all now tremendously wealthy.

Between that and the Snapchat pursuit, there were suddenly no more doubts about whether Instagram was worth $1 billion to Facebook. Instead, Systrom was getting constant questions—from the media, from his peers in the industry, from everyone—about whether he'd sold too soon.

THE NEW CELEBRITY

*"There are plenty of products that are iconic. Coca-Cola
is iconic. Instagram isn't just iconic. It's a phenomenon."*

—GUY OSEARY, MANAGER
FOR MADONNA AND U2

In late 2012, Charles Porch paid a visit to Randi Zuckerberg, Mark's older sister.

Porch, who managed Facebook's relationships with top celebrities, needed job advice. Should he try to join the tiny Instagram team, newly settled into the garage room at Facebook's headquarters? Instagram only had 80 million registered users, compared to Facebook's 1 billion. But already, he felt they could become the top destination for pop culture on the internet.

As they lounged in the grassy backyard of her 6,000-square-foot Los Altos home, drinking rosé, the question opened up old frustrations.

Randi Zuckerberg had been one of the earliest employees at Facebook. Ever since 2009, when President Barack Obama's administration decided Twitter would be one of his primary ways to communicate with U.S. citizens, she'd wondered whether it was possible for Facebook to have a similar role in the world. Could her younger brother's website be one that celebrities and musical artists and even presidents prioritized when they talked to their audiences? On top of her regular responsibilities as the head of consumer marketing, she developed a strategy to get famous people to post more.

But her plan faced two nearly insurmountable barriers: the big names were not very interested, and neither was Facebook. In the fall of 2011, a few months after hiring Porch, she resigned.

Randi Zuckerberg, a five-foot-five brunette, was as effusive and quirky as Mark Zuckerberg was robotic. Her dining room was decorated in purple wallpaper peppered with giant red lips, while around the table were a number of chairs in a variety of sizes and styles. She enjoyed public speaking and had grown up thinking she would be an opera singer.

She'd hired Porch from Ning, a firm that created mini social networks for celebrities' fans, back in 2010. Porch, a pale balding man with a gap between his two front teeth and a disarming manner, had an encyclopedic memory of names and faces and how celebrity networks worked. Well before the term "influencer" was in anyone's lexicon, he knew who you would need to lunch with in Los Angeles if you wanted all the famous moms to use a new Facebook feature.

Together, Zuckerberg and Porch experimented with every flavor of event featuring public figures on the social network, flying to more than a dozen cities while Zuckerberg was pregnant with her first child. Would it draw an audience on Facebook if Bono broadcast live from the World Economic Forum? What about if CNN anchor Christiane Amanpour did a video about the Arab Spring? Maybe they needed to represent Facebook at the Golden Globes? Go live with singer Katy Perry?

They did all of these things. Facebookers thought the strategy was frivolous nepotism—the CEO's sister spending company money to go cavort with famous people. At a company with engineers at the top of the hierarchy, it wasn't clear how these collaborations contributed directly to growth.

Lured by the Zuckerberg name, the celebrities took the meetings. But they were overwhelmed by Facebook and its fan page mechanics and likes and algorithms and promoted posts—so they tended to have staff running their accounts.

At one point, members of the band Linkin Park confessed they didn't even know if they had the rights to play their own music in a Facebook video, because they didn't know who had the deal to manage their fan page. In another instance, William Adams, better known as will.i.am of the Black Eyed Peas, got up during a meeting with Randi Zuckerberg and Porch and walked around the conference room playing games on his cell phone as they continued their pitch. *We're banging our heads against walls,* Zuckerberg thought.

She left the company before its initial public offering, after six years working there. Porch continued pushing, giving stars as big as Rihanna tours of Facebook's headquarters, without achieving much buzz. People were going to Facebook to talk to their friends and families and share links, not to keep up with celebrities.

Back on the Los Altos lawn in 2012, a couple glasses in, they resolved it together: Instagram was the right move for Porch. Despite skepticism from many at her brother's company, she had been right about celebrities, and how their participation could help cement a product's hold on popular culture.

There were some clues that Instagram was a promising place to apply the strategy. The stars who did have the app were managing their own accounts there, instead of hiring teams to do it for them. The network didn't require a clunky fan page like Facebook or an insightful 140-character comment like Twitter. Celebrities could post a simple square photo and immediately reach everyone they needed to reach.

Zuckerberg was more right than they realized. Instagram would grow beyond its initial roots as a creative space for photographers and artisans. It would metamorphize into a tool for crafting and capitalizing on a public image, not just for famous figures but for everybody. Every Instagram account would have the chance to be not just a window into someone's lived experience—as the founders initially intended—but also their individual media operation. The shift would birth an economy of influence, with all of the interconnected Instagram activity at its nexus, in territory uncharted by Facebook or Twitter.

Getting there—into this uncharted territory—started with Porch behind the scenes, influencing the soon-to-be influencers, hand-holding, and strategizing over many more glasses of wine.

· · · · · · · · · · · · ·

Porch, the gay son of a French mother and an American father, grew up speaking both French and English as his family split time between the countries. Because of his sister, who couldn't communicate verbally due to a disability, he learned to read expressions and emotions well. He absorbed strategy lessons from his father, who taught military history. There was no cable at home, so the family listened to classical music. Surprisingly, considering his later adventures in Hollywood, he'd had little exposure to pop culture. For three intense years, Porch was in choir school in Princeton, New Jersey.

He majored in international development at McGill University in Montreal, thinking he would become a diplomat. But he was drawn back to music—quite different than the kind he grew up with. He moved to Los Angeles in 2003 and found an internship via Craigslist at Warner Bros. Records, where he was tasked with finding ways to hype new music albums on online message boards from stars including Madonna, the Red Hot Chili Peppers, and Neil Young.

One of the Warner assistants was Erin Foster, the daughter of Canadian producer and songwriter David Foster, who had just worked on albums for Josh Groban and Michael Bublé. Because of her father, Foster wasn't given much work, and out of boredom, she kept trying to get Porch to sneak out to the Starbucks across the street. He usually protested, wanting to make a good impression at the internship. But with time, they bonded and became best friends.

The Fosters, who were well connected to Hollywood through music deals and related to the Jenner-Kardashians through one of David Foster's marriages, became like a second family to Porch, but in a completely different universe than his biological family, which had settled in a coastal town six hours north.

"As I brought him around people in my life, whether it was celebrities or well-known families, Charles was just never really fazed by anything," Foster remembers. Cycling through bad boyfriends, she had a dramatic personal life and relied on his stability. "He makes people comfortable because he's comfortable. And I think that he's intuitive to people's needs and wants."

Over the years, at Warner and then at Ning, and through the Fosters' friends, Porch found it was important to develop trust with celebrities through helping them navigate the confusing new digital frontier—not just by pitching a product. This was well before celebrities knew they needed digital strategies. He would discuss building online fan bases with Zooey Deschanel, Jessica Alba, or Harry Styles before it was his or anyone's job to do so. During his next job, at Ning, he would sign up just as many stars for Twitter, where he didn't work, after listening to people talk about what they wanted to accomplish.

By the time he was at Facebook, Porch had developed a theory about what would attract public figures to social media sites. He would find a way to talk to celebrities directly and personally about their goals, rather than going through their record labels or managers. He knew how to make their posts online sound natural and personal. If celebrities lifted the curtain on some of their private thoughts and experiences, they would build a bond with their fan bases. The online conversation would put celebs in control of their own brands, increasing their relevance and therefore their commercial potential.

· · · · · · · · · · · ·

A few days after talking to Randi Zuckerberg, Charles Porch walked over to Kevin Systrom's desk and explained his plan. He would go after the top users on Twitter and YouTube, then try to transition them over to posting photos on Instagram. And he would meanwhile ensure that Instagram's homegrown stars—the ones who were cropping up from the suggested user list and otherwise—got more direct support from the company.

Porch already had a wish list of people he hoped would eventually have accounts on Instagram, from Oprah Winfrey to Miley Cyrus. Once the stars understood what they could do with it, their audiences would follow, the way they had for Selena Gomez and Justin Bieber. And then more stars would follow their industry's leaders onto the platform, and more fans, and on and on. The public figures needed Instagram, and Instagram needed them—or at least, that was the pitch.

Systrom hadn't heard of Porch and was pleasantly surprised by his enthusiasm. The CEO was initially hesitant to embrace celebrities on Instagram, thinking his app was more about what people experienced and saw than about self-promotion. But he did realize that the community would need to evolve as it grew, and that if it was going to happen anyway, Instagram could have a hand in shaping it. He had always admired those who were the best at what they did, from high-end chefs to electronic DJs. He wasn't as well versed in mainstream pop culture, but Porch could fix that.

Systrom and Amy Cole, the business lead, were already on call to help a smattering of big names, from LeBron James to Taylor Swift, and they definitely needed someone else to take charge of that kind of work. As long as celebrities' posts weren't too promotional, they could give Instagram users a lens into previously inaccessible worlds, the same way Instagram had brought users behind the scenes with reindeer herders and latte artists. Celebrities managed communities just like Instagram did, and could help bring their fans to the app.

Porch thought the fashion community would be key to hitting the intersection of Instagrammy visual culture and mainstream culture. The fashion bloggers and models were already on the app, so Instagram would just need to convince the Anna Wintours of the world to take them seriously. Once fashion was onboard, the big names in Hollywood would be too. Then musicians would follow. And then so would sports stars. All the public-facing industries were connected, he explained.

Porch's first test project was at New York Fashion Week in February 2013. The night before the event started, he set up a basic presence for Instagram in the event's showcase tent at Lincoln Center, plugging in two screens and marking the territory with a small wooden carving of the Instagram logo. If people took a picture there, it would display on those screens.

I really hope Amy Cole likes this, Porch thought, aware that the Instagrammers had a very specific vision for their brand's look and feel.

The next day, when he got to the tent, he saw a crowd swarming around the Instagram setup, getting excited to see their photos show up live. Event photo booths, at this point, were novel. Models, designers, and bloggers all seemed equally willing to participate in the Instagram-branded experience.

This was the exact moment Porch knew he was working with something big, that would catch on naturally if he could do a little pushing and hand-holding with the right people. His strategy was all about finding the trendsetters to work with first; he knew that after that, their Instagram success would create peer pressure for others.

In order for his plan to work, those key people needed to know the person in charge of Instagram, and trust him. They needed to feel like they were supporting someone they liked, who could answer their questions, so it felt less like a chore. It was lucky for Porch that Systrom, unlike Mark Zuckerberg, was willing to make hobnobbing a priority.

• • • • • • • • • • • •

Systrom and Porch made their first trip to Los Angeles in 2013 with a new feature to woo celebrities: verification. Instagram was shamelessly copying a feature that Twitter offered, placing blue checkmark badges next to accounts to certify that the person behind them was indeed who they said they were. The verification badge had started as a measure to protect against impersonation but quickly evolved into a status symbol. If you were verified on Twitter, you were important enough that someone might want to impersonate you.

At the time, if you wanted to get verified on Instagram, the only way you could do it was by knowing someone at Instagram. Like Facebook and Twitter, Instagram had no customer service system or phone numbers to call, giving extra incentive to meet actual humans who worked there. That made the verification badge special, giving an impression that it equaled Instagram's endorsement of the person's posts—even though that wasn't what Instagram had intended.

Ashton Kutcher, the actor, and Guy Oseary, Madonna's manager, had kept in touch with Systrom since the day they visited the company and considered investing in 2011, the same year Systrom had saved Kutcher and his ski trip buddies from the burning cabin. Now the relationship could be valuable to the pair's other friends. So Kutcher and Oseary agreed to host a party for Instagram on the outdoor patio of Oseary's Beverly Hills mansion, inviting dozens of people who were interested in meeting Systrom, including Harry Styles and the Jonas Brothers. Most of the people invited came without their handlers. Instagram pitched in for drinks and hors d'oeuvres. At one point, after lifting a utensil to *clink, clink, clink* against his glass, Systrom introduced himself as the CEO of Instagram, Porch by his side.

For the rest of the night, people came up to him, asking why they should use the app; those who did already explained how it made them feel. Some said Instagram allowed them to speak directly to their fans, but also their friends. Others voiced concern over seeing hateful comments on some of their posts. A few of the stars exchanged phone numbers with Systrom, promising to tell him if they ever needed help, or if they thought of something in the app that could be better.

While the music artists were familiar with selling themselves, movie stars were not. "Trying to sell Hollywood on why this would be valuable was pretty difficult," Kutcher remembers. "It's not great as an actor for people to know who you are as a person, because it makes it harder for them to imagine you as a character." But Kutcher thought it was inevitable that in the digital age even movie stars would have to stop being so mysterious, because eventually casting decisions would be swayed by the ability to bring an audience to a movie, like he could with his Twitter followers. "It seemed clear in the entertainment industry that there would come a day when people would be valued as entertainers based on their ability to sell the product they were in," Kutcher explained.

At first, Systrom felt out of place among these celebrities; the feeling reminded him of his days at Middlesex boarding school, where his peers had yachts and summer homes and families who were in the news. But as he asked more questions and heard their stories and learned of their insecurities, he realized that everyone at the party was just trying to be better at their jobs. They could help each other.

Instagram encouraged celebrities to use the app to document what they saw in their daily lives, taking power back from the paparazzi and controlling their own narratives. But stars posting on Instagram required a careful balance, different from what the paparazzi offered: if celebrities only logged on to post about their upcoming albums or movies, their followers would see their efforts as promotional. If they included that content with organic posts from their everyday lives, they would become relatable, and then their followers would be more likely to cheer for their commercial success.

Stars were used to being paid for their photos that showed up in celebrity magazines. But Instagram would not be compensating anyone—not directly, at least. Porch said his team was willing to offer advice on Instagram-related projects, to act as free consultants for those who knew the right number to call. *If you're not going to be good at Instagram, don't do it,* he would advise. The sentiment built trust—and intrigue. (Eventually, celebrities would learn to make money off their Instagram accounts, but the idea sounded tacky at the time.)

Oseary watched Systrom at the party and noticed that his way of working didn't seem transactional. Compared to others in the technology industry, he was easygoing, trying to be a friend more than a salesperson, working to genuinely understand the product's effect on high-profile users. It was hard to imagine Mark Zuckerberg ever mingling at a party like this, with his Secret Service–level security detail and his public relations entourage.

But while Systrom immersed himself in celebrity culture, he could also be clueless about it. A short brunette woman at the party explained that while she loved using Instagram, she thought it was pressuring young people, who could be quite mean to each other online. Because the stars had much bigger follower counts, the app's upsides and downsides stood out to the extreme. She could see her fans getting bullied in the comments for her photos—something Instagram didn't have a solution for.

"And what is it that you do?" Systrom asked, his six-foot-five frame hovering over hers. She took out her phone to show him her Instagram profile. She was a pop star with about 8 million followers, and her name was Ariana Grande.

• • • • • • • • • • • • •

Some celebrities didn't rely on Instagram's word about the value of the app, choosing instead to do their own research by asking early adopter peers for advice. Kris Jenner, the matriarch and business boss of the Kardashian-Jenner reality television family, fielded lots of phone calls from her high-society friends in 2013 and 2014. They asked why her daughters bothered with so much Instagram.

"A lot of people thought that without a level of privacy and mystery, they wouldn't be as interesting," Jenner explained. "For so many people in the entertainment business, the only way they wanted to share things was if they were doing a proper interview, or if they were doing a show on television."

Since the Kardashian-Jenner family had already shared so much of their lives on TV, they had no inhibitions about sharing them online. A couple years after their reality show started in 2007, their producer, Ryan Seacrest, called Jenner to suggest that her most famous daughter, Kim Kardashian, start talking to fans on Twitter. So she did, learning what worked and what didn't, and proceeded to teach the rest of the family.

In 2012, Kim Kardashian joined Instagram, wanting to repeat her Twitter success in a new market. Her audience was thrilled to have an opportunity to follow beyond the show while seeing more of her iconic and controversial hourglass figure. As members of the family accumulated millions of followers, Instagram became their main branding tool, eclipsing Twitter in importance because of the immediacy and intimacy the images provided.

When Porch and Systrom were hanging out in Hollywood and promising brand control for stars who used Instagram, they weren't explicitly mentioning to those stars the potential to derive extra income by posting about brands and products. But Kim Kardashian knew what was possible.

Kardashian had learned from socialite friend Paris Hilton in the early 2000s how to use photography to build a brand. Hilton had learned it from her manager at the time, Jason Moore, who had devised a complicated system of manipulating the media in a pre-Instagram world. Moore was the progenitor of the modern idea that someone could be *famous for being famous* and shamelessly build a business around it.

On the reality show *The Simple Life*, Hilton played the role of a blonde airhead rich girl—a personality she would say, years later, that was at least partially the invention of the show's producers. Either way, she was willing to go along with a plan to make the entire world—not just her *Simple Life* set—a stage. A leaked sex tape launched her into the tabloids, and there she stayed. Moore was tipping off the paparazzi as to her whereabouts, building relationships with trusted photographers and fueling an always-on celebrity news cycle, suddenly possible because of the rise of internet blogging. New media sites like *PerezHilton* and *TMZ* thrived on Hilton drama.

When Moore looked at Hilton, he saw his chance to mold a person into a brand—a new type of brand, unlike Oprah's media empire or the Olsen twins' merchandised acting career. In college, he'd spent a semester learning about the success of Mattel's Barbie doll. "I started thinking, If Barbie could walk and take a shit, what would she be like?" Moore recalled. "What would be her brand? Because right now Barbie is a lifestyle. She's a woman with a glamorous life, who lives in a good house and has great accessories. Why did that capture America and the world and youth?"

Moore tried to turn everything Hilton did into a moneymaking venture, even trademarking the phrase "That's hot!" which Hilton said often on the show, so they could put it on T-shirts. Soon Hilton had a perfume line, a clothing line, and philanthropic projects: she had turned "famous for being famous" into a new kind of entrepreneurship. In a society without social media or the iPhone to show fans' enthusiasm, Moore would bring his own camcorder around the world to create video reels of Hilton arriving in new cities and launching products, so they could edit and present the footage to potential business partners. By seeing Hilton around her enthusiastic fans, brands understood the value of attaching her name to their projects.

Hilton had money, so in moments they really needed to control her message, they would use it. Moore would pay a paparazzo to wear a green scarf so Hilton would know exactly whose lens to look into when she stepped out of her house, out of a club, or, at one point, out of jail. Then Moore would anonymously broker a deal to sell the picture to a celebrity news site. "Then the publication would come back to us and ask for a comment—and the whole time they had no clue we were behind it," Moore explained. "The paparazzi was essentially Paris's daily Instagram post, and that reality show was the weekly Instagram story."

Meanwhile, Kris Jenner realized that the fastest way to achieve fame was by being associated with more famous people (the concept would later help Jenner create mini stars on Instagram out of the stylists and trainers and makeup artists who worked with her family). So in 2006, before *Keeping Up with the Kardashians* began airing, she called Moore to ask if Hilton and Kim Kardashian could appear together more often, as her daughter was looking to build a clothing business called Dash. Kardashian was a much curvier brunette, who would appeal to a different kind of consumer entirely, Moore thought. He told Jenner it would not be a problem.

Hilton's carefully controlled images and videos fueled her business. So when digital platforms like YouTube and iTunes eventually reached out for the opportunity to feature Hilton's videos or music for nothing in return, Moore dismissed them. "We were used to getting paid hundreds of thousands of dollars per photo," he explained. "Why would we do that for free?"

But Jenner and Kardashian, who were still early in building their fame when Twitter launched, couldn't make as much money off leaked photos. They realized that they could create an even bigger business in social media, by building their own version of a Hilton-inspired lifestyle brand, then selling ads based on the audience, as Moore was doing manually.

Instead of leaking photos to the media, or paying paparazzi, or making reels for brands, they could release images of themselves on Instagram to an audience potentially much larger than the circulation of any pop culture magazine. Down the line, as they attached products to their fame, they could get feedback on what people wanted to buy before they even developed what they wanted to sell them. Kim Kardashian would ask her followers what color her perfume bottles should be, and get an answer via an instant vote.

But this dynamic was largely invisible to most celebrities who weren't yet on Instagram. Jenner remembers one conversation with an A-list celebrity who, like some of the attendees at the Oseary party, questioned the value of having followers at all. Jenner realized, "It's fun to be a part of the social dynamics of what's going on with Instagram, but at the same time if you *do* have a lot of followers and you *do* want to go into the business of selling something to your fans, it sure is an instant audience of people right there ready to join the party."

The Kardashians accepted hefty fees from brands for incorporating products into their posts, and, like Snoop Dogg in 2011, often failed to mention whether they made money doing so. The lack of disclosure made the posts feel less like ads and more like helpful tips—and U.S. regulators were slow to catch on to the practice.

Since consumers are much more likely to be swayed to buy something if friends or family recommend it, as opposed to advertisements or product reviews, these ambiguous paid posts were effective. The Kardashians, who built a fan base by being vulnerable on television and then on Instagram, were able to make their followers feel like the family was their friends, not salespeople profiting off their consumption. Their Instagram endorsements were so powerful, whatever they put their word behind would sell out quickly—whether it was makeup, clothing, or wellness products of ill repute, like their dieting teas and modern-day corsets called "waist trainers." The Kardashian empire on Instagram was like Oprah's Book Club in the late 1990s, with a supersize silicone injection.

Influencers like the Kardashians helped brands get around the pitfalls of online commerce. With the rise of Amazon and other sites, consumers had an abundance of options for whatever they wanted to buy. Before making purchases, they would spend time reading reviews or hunting for the best deal. Branded posts on Instagram provided a rare opportunity to get consumers to make a spontaneous decision, since a trusted person's endorsement made them feel they were making an informed choice, even about products as dubious as waist trainers.

Today, Kim Kardashian West has 157 million followers and makes about $1 million for a single post. Paris Hilton eventually joined Instagram too, and now has 11 million followers. Porch now has employee counterparts in Los Angeles who answer celebrity queries for the Kardashians and others, solving their problems directly while most of the app's users fend for themselves.

· · · · · · · · · · · ·

Years later, as millions more people became Insta-famous enough to post sponsored content, perusing the accounts of the Instagram elite would start to feel like visiting an alternate reality where anything negative in life could be cured by a purchase. There would be semifamous people pretending to be vulnerable so they could sell products that they pretended to love, which supported a lifestyle they pretended was authentic. The flurry of aspirational branded posts would manipulate the masses into feeling bad about their normal lives. The effect would depress some of the early Instagram employees, who had wanted so badly to build a community centered around the appreciation of art and creativity, and instead felt that they had built a mall.

But that future would come into focus only after enough Instagrammers had built up their fame. And back in 2013, giving people the chance to build followings on Instagram seemed wonderful and powerful. Instagram wasn't just for celebrities—it was for everyone. Employees thought of the app as a democratizing force, allowing regular people to bypass the normal societal gatekeepers and simply show, based on their Instagram following, that they were worth investing in. The Instagram follower number became like a Q score, a way of measuring brand recognition for anyone—whether they were known for their travel photography, their baked goods, their pottery, or their fitness routines.

Building a following on Instagram worked differently than on other apps. Because Instagram had no share button, people didn't become famous there by going viral the way they did on Twitter. No one could re-share content someone else had posted. New employees of Instagram, especially those coming from Facebook, would regularly suggest sharing tools to help increase the amount of posts on the app, only to be shot down by Systrom and Krieger. Public re-sharing was such a popular request that other entrepreneurs built apps like Regrann and Repost to attempt to fill the need, but these were no substitute for an in-app function. This made it harder to get noticed, but in some ways made it easier to build a personal brand. All your posts were yours. That was what the founders wanted.

There were some ways to manipulate the system. There was still a "Popular" page, showing what was trending on the app. There were hashtags, via which people could discover others they weren't following. But by not allowing automated virality, Instagram was still exerting some control over who became famous or not.

The community team's efforts, originally meant to highlight interesting content that could serve as a model for new users of the app, had a side effect of forcing these interesting users into the limelight. The team handpicked Instagram handles to share with the wider Instagram community, determining not just *what* would become popular on the app, but also *who*. And as the number of people using the product grew, so did the team's power.

Porch saw the community team's kingmaking as an opportunity to embrace. For Instagram to be an aspirational place not just for celebrities but for everyone to post photos, it had to be unique, with its own homegrown trends and personalities. And it would be up to Instagram to support new stars, not with money directly, but with attention and opportunity.

And so, as more people built up audiences for their Instagram accounts, the biggest influencer of the bunch was Instagram itself. Most Instagram users were ordinary people, without connections to big businesses or celebrities whom they could ask to mention them in a post to boost their visibility on the app. But these regular people could get an immediate boost from Instagram's own curation tools: the suggested user list and the @instagram account, which had more followers than any celebrity.

. . . . • . . . . .

The community team specialized in discovering users who were becoming prominent in specific categories, like fashion and music. Dan Toffey, for instance, was the Instagram employee focused on discovering pets. He kept a running spreadsheet of the best pet accounts, trying to be as unbiased and equitable as possible. The list consisted of cats, dogs, bunnies, snakes, and birds, some adopted, some expensive purebreds, some scraggly, some immaculately groomed. He would parse through his list to select ones for a feature called "The Weekly Fluff," where he would showcase the accounts doing great work on the @instagram page, in hopes that they could inspire others.

Professionalism aside, Toffey most appreciated animals that looked goofy, in need of a little extra love. Baby goats missing their hind legs that got around on little wheelchairs, for instance, or cats with their tongues permanently sticking out. But especially tragic dogs. An awkward-looking Chihuahua-dachshund mix caught his attention, with its elongated snout and overbite.

The dog was named Tuna, and its owner, Courtney Dasher, an interior designer, had adopted him at a farmer's market in 2010 when she saw him, toothless and shivering, in an oversize sweatshirt. When Dasher joined Instagram the next year, she decided to show Tuna's face instead of hers, on an account called @tunameltsmyheart. The dog account's popularity spread beyond her family and friends to a few thousand people. But on a Monday night in December 2012, the account started gaining fans around the world.

After Toffey posted three pictures of Tuna on the Instagram blog that night, the dog's following grew from 8,500 to 15,000 within 30 minutes. Dasher pulled to refresh the page: 16,000. By the next morning, Tuna was at 32,000 followers. Dasher's phone started ringing with media requests from around the world. Anderson Cooper's talk show offered to fly her to DC; she appeared via webcast, thinking it wouldn't be feasible to take a vacation day.

But as requests for appearances continued to come in, her friends warned her about what was coming before she realized it: she would have to quit her job at the Pacific Design Center in Los Angeles and run her dog's account full-time. It sounded ridiculous, so she took a month off to test the theory. Sure enough, BarkBox, which made a subscription box for pet items, was willing to sponsor Dasher and her friend on an eight-city tour with Tuna.

People in various cities came up to her, crying, telling her they were struggling with depression or anxiety and that Tuna was bringing them joy. "That was the first time that I realized how much weight these posts had for people," Dasher later recalled. "And that's also when I realized I wanted to do this full-time." Her life became about managing Tuna's fame.

Berkley, part of Penguin Random House, signed her up to write a book titled *Tuna Melts My Heart: The Underdog with the Overbite*. That led to more brand deals, plus merchandising to put Tuna's likeness on stuffed animals and mugs. In her book's acknowledgments, she thanks Tuna most of all, but also Toffey for sharing the post that changed her life. The tastes of one Instagram employee directly affected her financial success, but also the habits of the two million people who now follow that dog—including Ariana Grande.

· · · · · · · · · · · ·

Most people never knew the Instagram employees who shaped their careers. Marion Payr joined at the suggestion of her husband, Raffael, who saw the app featured in a magazine in 2011. She was just using the app to share photos of her travels. The Austrian woman in her thirties was working a desk job in the marketing department of a television company in Vienna and had no prior experience in photography. One day in 2012 she received an automated email from Instagram, telling her she'd been selected as a suggested user. Followers for her @ladyvenom account ballooned from 600 to many thousands.

Payr decided to embrace the miniature dose of fame, befriending others on the list around the globe, all of them puzzled but many of them grateful. Soon she was going on local photo walks and helping organize InstaMeets, becoming a volunteer ambassador for the company in her country, even though she'd never met or corresponded with an Instagram employee.

Eventually, she was able to quit her job and devote herself to travel photography full-time. She built a small studio to consult with brands on the side about how to use Instagram strategically. Once she had 200,000 followers, all of them wanted to build an audience like hers. She was considered an expert in getting attention on the now-lucrative Instagram app, but still had no idea why she'd gotten popular.

· · · · · · · · · · · · ·

Though it seemed like getting featured by Instagram was the ideal outcome, not everyone enjoyed feeling like they had a sudden commitment to entertain thousands of strangers. Some who were featured felt beholden to their new audiences for a while and then quit, overwhelmed by the pressure. It was like winning the lottery: worth celebrating but complicated, and how to cash in was not as immediately obvious.

Still, bloggers tried to decode the process for getting featured by Instagram, via the company's posts on its @instagram account or the suggested user list. There was little to explain, because no formula or algorithm existed. In contrast to Facebook's decisions, which were data-driven, Instagram's curation developed out of its employees' personal tastes.

And what Instagram gave, it could take away. People got booted from the suggested user list, for example, or had their accounts canceled without warning or explanation for violating ambiguous content rules. Few people realized that choosing to build a business on Instagram meant placing one's future at the mercy of a small handful of people in Menlo Park, California, making decisions on the fly. The only way to be certain nothing bad would happen was to build a relationship with an Instagram employee like Porch or Toffey. As Facebook would say, the strategy didn't scale.

Instagram employees disliked their one automated machination of buzz, the "Popular" page, which circulated posts that got a higher-than-average number of likes and comments. The company would eventually eliminate it. With no human tastemakers in the way, the page was easier to game, the way Twitter and Facebook could be gamed. Those who were trying to build an audience learned to post at ideal times of day, like during lunchtime, or in the late afternoon or late evening, when people were most likely to be checking the app. Once they succeeded in landing on the page, they would gain more followers, making their next posts more likely to succeed. People pursued higher metrics, not realizing what they could do with followers and attention until they had it.

Paige Hathaway was one of the earliest to benefit from the Popular page. In 2012, the then 24-year-old started posting photos on Instagram charting her progress of getting in better shape. She was a thin blonde, recruited by a trainer she had met at the gym to compete in a body transformation competition to become more muscular.

At the gym, people were puzzled to see a sweaty person taking pictures in the mirror with her phone, as if working out was at all glamorous. But on Instagram, strangers found it interesting to watch an attractive woman become more toned. Over the summer of 2012, she went from 100 pounds to 120, building her strength and a chance at winning prizes in the competition she was in, coming in second place.

Becoming better at fitness was a way to gain control over her life and her future. Hathaway had jumped from home to home in the foster care system during her childhood. Then she'd taken on various jobs to put herself through Oklahoma University. After she started working out, she explained, her "confidence levels were at a level they were never at before." She became a personal trainer and kept posting, even without a specific competition to work toward.

Hathaway wasn't sure what kind of career she wanted to have long-term, but her posts started appearing on Instagram's Popular page every couple weeks. Building an audience brought opportunities to her, before she knew to ask for them. "I was getting all these companies reaching out wanting to work with me and I had no idea what that meant," she remembers. She decided on being the face of Shredz, a small bodybuild-ing and weight loss supplement company, when she had a mere 8,000 fol-lowers. Once Instagram added the ability to upload video in the summer of 2013, the app became ideal for demonstrating workout moves. Hath-away's followers skyrocketed into the millions, and so did her income. Shredz too followed alongside her, becoming a multimillion-dollar com-pany. Sweaty Instagram pictures in the gym mirror became acceptable workout behavior.

"I had to hire help," Hathaway says of her quick rise. "The first two years I had an entire team behind me. I hired a management team, I had people helping me with clients online, I had people helping me with en-dorsements. It was beyond my own self to manage everything."

Her success shocked the fitness industry, raising questions about what a star in bodybuilding should look like. Hathaway had gotten her audience—and lots of traditional media attention—without paying any of the normal dues, like participating in fitness competitions. In early 2014, Arvin Lal, the CEO of Shredz, had to defend his decision to use Ha-thaway instead of a career fitness competitor for his product marketing: "Who's to say the person onstage has better fitness or body than a person with a million Instagram follows? Paige is probably the largest female fitness model across the world. Marketing and being able to touch people *off* the stage is more important than being able to touch people *on* the stage."

• • • • • • • • • • • •

Several industries were undergoing the same overhauls and grappling with similar questions. If something became popular on Instagram—whether a fitness routine, a home decor trend, or a flavor of cookie—did that make it more valuable in real life? Were endorsements from the Insta-famous worth seeking out or paying for? And if you had a popular brand in real life, should you try to get popular on Instagram too?

As the chief creative officer of Burberry in London, Christopher Bai-ley would routinely make trips to Silicon Valley to get more ideas about how to produce a modern fashion brand. Ahead of a new iPhone release, for example, Burberry collaborated with Apple, working under extreme nondisclosure agreements to come up with something to say about phone photography at the launch of the iPhone 5S in 2013.

On one of these trips, after Porch's outreach at New York Fashion Week, Bailey met with Systrom and was inspired by his vision for Instagram as a source of behind-the-scenes moments. He started to notice accounts documenting fashion in the street, some of it Burberry, and was struck by how quickly new fashions appeared on the scene and were discussed by prominent accounts. The Instagram users wouldn't wait for whatever print advertising was planned on Burberry's calendar. Bailey realized that Burberry would need to start posting its own content to get ahead of the coming industry transformation.

"We were used to going through a very long laborious process of organizing shoots and productions on those shoots, and a classic media buying program working with magazines," Bailey explained. "Six or nine months later you finally saw that image in a magazine. With Instagram, the fact that we could hire our own photographers, our own team, and within minutes it could be online and we would have dialogue directly with people who were interested in our brand, was just incredible."

In September, around the iPhone launch, Burberry invited Systrom and a few other Instagram employees to its fashion show in London. Bailey hypothesized that the future of fashion events wouldn't just be about the runway walks and styles, but also about bringing a wider audience into the scene—telling them who was wearing those styles, who had attended the event, and whether the whole experience was memorable and worth following on Instagram.

So on Burberry's runway, they changed the way they had always done fashion shows, playing music for the first time, and inviting nonprofessional photographers, specifically Instagram street-style photographers, to document the show as it happened on their new iPhones, provided by Apple. These amateurs were allowed to post without explicit approval from Burberry. And Bailey had to make sure the skeptics understood why he was doing it. He recalled that "there was a lot of cynicism in our industry about what we were doing with Instagram, and comments that luxury

customers would never use this kind of platform, because it was too ubiquitous. Before that, fashion brands had been kind of sacred, veiled in
secrecy. We would push out polished images of what we wanted people
to see."

This switch was risky. Bailey spent a lot of time in internal meetings
explaining how hashtags worked and why it was okay to have negative
customer opinions appear alongside positive ones on Burberry's Instagram posts. He argued that the brand couldn't avoid Burberry's presence
on Instagram, whether the fashion house was participating or not, since
regular people were talking about the brand with a #burberry hashtag
regardless, so they might as well be part of it.

Bailey didn't have to defend his Instagram-inspired strategy for long.
A month after the runway event, Bailey's boss, Angela Ahrendts, left to
be an executive at Apple. Soon after, Bailey was promoted to CEO.

· · · · · · · · · · · ·

The iPhones that Burberry celebrated that year included software directly influenced by Instagram. For the first time, iPhones offered a way
to take photos in square format, so they would be ready to post on the app
without having to be configured or edited. Apple also added some of its
own filters directly into its camera tool.

The fact that Instagram's growth was unthreatened by this move
was perhaps the clearest indication that Instagram itself was no longer
about sharing filtered photos on other services, or about filters at all. Its
power was less about technology, and more about culture and networking, thanks to the team's outreach and curation since the app's earliest
days.

When Systrom and Porch traveled to London in 2013—making their
first trip to promote Instagram internationally—they combined public
figure strategy with community strategy. They attended not only Burberry's runway show, but also a meal hosted by chef Jamie Oliver. Oliver
had been one of the first celebrities to sign up for Instagram, long before
the Facebook acquisition. Systrom, introduced to the chef by an investor, had nervously created an account for him at a dinner.

In 2013, like Oseary and Kutcher in Los Angeles, Oliver was able to pull together an impressive set of London stars from movies, music, and sports. The actress Anna Kendrick attended his meal event, as did members of the Rolling Stones and the cyclist Chris Froome. That same night, Instagram hosted an InstaMeet at London's National Portrait Gallery with some of the high-powered community members who weren't otherwise famous. As usual, Systrom was asking questions, gathering feedback, and making contacts.

What Systrom and Porch did on this trip became a formula for future ones. They'd have at least one intimate meal with celebrities, one event with regular users, and one public moment, like a fashion show or a soccer game.

●　●　●　●　●　●　●　●　●　●　●　●

Instagram was making more headway with public figures, the area that Twitter had historically dominated in social media, just as Twitter was preparing to go public. Nobody knew how valuable Wall Street would consider Twitter to be, or whether it would end up a formidable competitor to Facebook in the eyes of investors. Mark Zuckerberg, always fiercely competitive, wasn't going to take any chances.

Facebook, two years after Randi Zuckerberg's resignation, finally had a reason to start courting public figures to the bigger social media site. The move could inflict pain on Twitter. In the lead-up to Twitter's IPO, Facebook spent months doing what Randi had always hoped they would. They built out a global partnerships team, tasked with recruiting posts from public figures.

Facebook's strategy was different than Instagram's, tailored more to relationships at institutional levels—with record labels, TV studios, and talent agencies—as opposed to Porch's strategy to forge direct relationships with stars. Facebook also reached out to media organizations like the *New York Times* and CNN, hoping the social network could provide an alternative to Twitter for posting important news stories. Facebook started allowing news websites to embed public posts in their articles, the way they could with tweets. The news organizations were willing to take cash incentives from Facebook to experiment on the site, as their traditional revenue streams like print subscriptions were drying up.

Mark Zuckerberg started referring to any Twitteresque posts on Facebook as "public content," and started saying on earnings calls with investors that he wanted to make this type of posting a priority for the company. He wanted Facebook to be better at Twitter than Twitter was.

A bonus of this strategy was it gave people more things to post and talk about on Facebook. With every year people were on Facebook, they were broadening their friend networks. It turned out that even if "connecting the world" was a great business objective, synonymous with growth, the side effect was that everyone's feed was full of loose acquaintances. Almost a decade after the company's founding, Facebook users were less willing to post about their more intimate observations and life events for this wider audience. Facebook was still adding users and revenue at an impressive clip—but Mark Zuckerberg liked to look at growth problems looming around the corner, and solve them before they became serious.

Facebook theorized that celebrity content and news could be a good conversation starter with people their users knew less well, or used to know in a different era of their lives. It could also generate data on users' interests, which would help lead to more accurate Facebook ads.

Instagram was operating so separately, Facebook barely considered it part of the strategy. Whatever progress Instagram made would barely count, unless it helped Facebook too. But they would collaborate. The Instagram team was small, so they'd ask Facebook for introductions in countries they didn't have staff. Other times, Facebook would lean on Instagram's relationships, encouraging celebrities, when posting on Instagram, to click the option to allow the post to appear on Facebook simultaneously.

For that, Porch was a big help. He convinced Channing Tatum that selling pictures of his new baby, Everly, to glossy celeb magazines was tacky. Instead, he argued, Tatum should post the first picture on Instagram and cross-post it to Facebook, explaining that the choice would appear innovative. Tatum agreed, and the post got more than 200,000 likes—with plenty of media coverage to boot.

Celebrities were often confused by the two products being part of the same organization but having different rules and strategies. Facebook, unlike Instagram and Twitter, was willing to dole out incentives to encourage celebrities and media organizations to create the kind of content they wanted. The main currency they used to reward such behavior with public figures was not straight cash, but ad credits—tens or hundreds of thousands of dollars in free advertising on Facebook. Tatum received some in exchange for the baby post, to promote an upcoming movie, but only because he posted on Facebook too, not just Instagram. That was more valuable than what the celebrity magazines would provide.

Tatum was something of a trailblazer, but soon celebrities would be posting life events on Instagram without Porch needing to convince them.

• • • • • • • • • • • •

Zuckerberg was worrying about Twitter competition more than he needed to. Since Facebook had been the first social media company to go public, the company trained Wall Street on the appropriate valuation model—the one that benefited Facebook most, that every single move it made was focused on driving. Facebook's strategy wasn't about buzz. It was all about growth.

By the end of 2013, Facebook was making about half its advertising revenue off mobile phones—dramatic progress in just more than a year, thanks to Zuckerberg's forced laser focus on solving the issue. The social network had 1.1 billion users. Zuckerberg had proved his thesis that wherever there was a growing network, an advertising business would follow. Facebook's stock was trading at around $50 in December 2013, up 80 percent since the beginning of the year, far beyond the $38 IPO price. Wall Street, used to modeling the future based on comparable examples from the past, was hungry for the next Facebook. Twitter was supposed to be the one.

Dick Costolo, Twitter's CEO, knew they couldn't beat Facebook at the growth game. But when preparing filings for the Securities and Exchange Commission, he realized they would have to provide the same "monthly active users" metric as Facebook, even though he could see a slowdown coming in future quarters. Twitter just hadn't been as focused on growth as Facebook was, and didn't have any alternative metric to give about world impact or importance. Plus, the SEC would probably require something comparable.

Twitter went public in December 2013 at $26 a share, and on its first day trading was already up to $44.90. By the end of the month, the stock had peaked at $74.73, revealing the market's tremendous optimism after seeing Facebook recover from its troubled debut. Twitter had about a fifth as many users as Facebook, and all the hype assumed they would eventually come closer to the same size, over time.

A couple months later, the company's first earnings report came out. Costolo thought it would be well received, since they'd sold a lot more ads than anyone expected.

He was wrong. Investors fixated on the slowdown in user growth, which he hadn't expected them to care about so quickly. The investors realized that if revenue growth follows user growth, the opposite must also be true—that any slowdown in user growth would lead to a slowdown in revenue growth.

Twitter's real strengths were difficult to explain. What was the value of being the place where all the biggest names in politics, media, and sports talked about all of the things the public cared about, before they talked about them anywhere else?

Even if Wall Street didn't understand, Instagram did.

• • • • • • • • • • • •

Facebook was the app to beat in terms of financial success and size, and Twitter was the one to beat in terms of cultural impact. Instagram was still the underdog compared to these two, if considered separately from its parent company. It was just starting to try advertising, with a quarter of Facebook's users and a handful of public figures using the site. But its strategy was quite different. Instagram, with no virality, was focused on training and curating content that could serve as examples to other users, getting celebrities to share the behind-the-scenes details of their lives. Twitter was based on live events and virality, so they wanted stars to use the site to do things that would start conversations and lead to a lot of retweets. Nowhere was that more obvious than at the March 2014 Oscars.

Twitter's television partnerships group had spent months with host Ellen DeGeneres's team, tossing back and forth ideas about how she could create a tweetable moment from the star-studded awards event. DeGeneres liked the idea of taking a selfie. Selfies had boomed in popularity since Apple had introduced a front-facing camera in its devices and since Instagram had popularized social photography. "Selfie" had even been the *Oxford English Dictionary*'s word of the year in 2013.

During rehearsal, DeGeneres saw a seat labeled with Meryl Streep's name, near the aisle in the third row. It gave her the idea that if she could get Streep involved, her selfie would be even more exciting. Representatives from Samsung, a major sponsor for the Oscars, watched her practice her lines and heard her mention the plan. They jumped on the opportunity, calling an ad executive at Twitter to ensure that if DeGeneres posted, she wouldn't use her personal iPhone and instead a Samsung one. The team presented her with a tray of Samsung options the morning of the event, all selfie-ready.

While live, Oscars host DeGeneres stepped off the stage and walked over to Meryl Streep. Bradley Cooper, also in the audience but unaware of the plan, improvised, taking the phone from the host's hand and bringing other actors into the shot: Jennifer Lawrence, Lupita Nyong'o, Peter Nyong'o, Angelina Jolie, Brad Pitt, Jared Leto, Julia Roberts, and Kevin Spacey. The resulting post immediately became Twitter's most popular of all time, retweeted by more than 3 million people.

· · · · · · · · · · · ·

When the team at Instagram saw what Twitter had accomplished at the Oscars, and all the media buzz it generated, they were frustrated. They didn't have viral sharing, but they were doing mini versions of these coordinated celebrity moments all the time. And they weren't just building relationships with top users; they were curating and promoting content from an entire ecosystem of interesting people, some of whom were becoming mini celebrities in their own right.

Instagram again had a chance to beat Facebook's competition in a different way than Facebook could. For better or worse, the app had become the perfect place for seemingly spontaneous moments that had actually been coordinated by corporate branding teams over months. Even without the company's help, brands were finding value on Instagram, fueled by outside advertising dollars and a growing crop of users who realized they could make a living on the app, like Paige Hathaway with fitness and Courtney Dasher with her dog Tuna.

It was during this more commercial, strategic phase of Instagram that the company's first employee, community team architect Joshua Riedel, decided to leave to get his master of fine arts degree in creative writing. Bailey Richardson, one of Riedel's early pre-acquisition hires who'd found those first photographers, artists, and athletes to add to the suggested user list, also decided it was time to move on. The artsy, magical novelty of Instagram's early days was fading with its size. And meanwhile, the original employees were far outnumbered by Facebook transplants and new hires.

Systrom told employees they were now dealing with not just one user community, but several, and they couldn't be good at reaching all of them. So they'd need to choose. He argued that besides mainstream celebrities, Instagram needed to use its limited resources to cultivate relationships with a few types of users—in fashion, photography, music, and teen groups—really well. They were not immediately going to prioritize relationships in food, travel, home design, or any of the other industries being shaped by the app's popularity, because any outreach would signal a commitment for the long term, and they didn't want to make promises they couldn't keep.

Before his departure, Riedel tried to pick people who would be good at improving Instagram's relationships with users in all of the priority categories—not tech people, but people who were part of the worlds they'd be tasked with reaching, full of sincerity. He hired people like Andrew Owen, who ran an annual photography festival, as well as Pamela Chen from *National Geographic*, to convince skeptical photographers and artists that Instagram was a legitimate place to put their work. Kristen Joy

Watts came in from a creative agency to focus on fashion, to cultivate an already-excited user base. He also hired Liz Perle from the *Huffington Post* to focus on young people, especially teens, who would be key to the future of the app. Employees like this were insurance, tasked with keeping the community positive and recognizing up-and-coming accounts that could be examples for the rest of the users.

David Swain, the Instagram communications chief, had two things he liked to say about the media strategy. One was "extend the honeymoon": keep people feeling good about Instagram for as long as possible in the wake of the Facebook marriage. The other was "don't fuck it up": avoid losing user trust, the way Facebook had. In order to achieve these goals, he thought, Instagram needed most of the press about Instagram to be about its best users, not about the company itself. And Instagram needed to remain behind the scenes for as long as possible. It was like Instagram was running a constant influencer campaign, for Instagram.

Swain was a Facebook veteran who'd joined that company's communications team in 2008 and had helped it weather several public crises. He understood what it meant to try to explain shifts in company strategy to an untrusting public. Right before joining Instagram in 2013, he was managing Facebook's PR around its relationship with outside game developers, who were building businesses on top of Facebook users' friend networks. (This open data-sharing with the developers would, in 2018, get Facebook in trouble with regulators worldwide.)

People had no such misgivings about Instagram, so Swain wanted to get out ahead and reinforce all the good things that were happening on the app, in a way that would feel natural and helpful, like it hadn't been Instagram's idea.

The communications team focused on making reporters' lives easier. Swain would meet up with journalists to explain how to understand trends and events on Instagram. Porch personally set up a touchscreen for *E! News*, so the celebrity channel could more easily discuss noteworthy Instagram posts, as an alternative to Twitter in their coverage. Liz Bourgeois, also on the comms team, would pitch the media stories about Instagram trends. Most users knew about the common hashtags on Instagram, like

#nofilter for a photo that was authentic and unedited, or #tbt for Throwback Thursday, for posting a photo from the past. Bourgeois would try to get the media interested in new hashtags like #catband. In that corner of Instagram, people were posing their felines with instruments to look like they were playing music.

If reporters for magazines and blogs would ask Instagram for some of the best accounts to follow in a particular country or industry, community team members would send along names to feature in slideshows or lists of the "Top 10 Instagram Accounts in London," or "Best New Fashion Photographers to Follow on Instagram."

It was tricky work, because whoever they mentioned would have their account more easily discovered via Google search, and was therefore more likely to get picked up by brands to do paid promotions. The Instagram employees didn't want it to be known they were selecting favorites, making some users' careers over others.

Despite efforts to market Instagram through its model users, the company still didn't publicly approve of them accepting money to promote products. Brands overall were paying about $100 million—an experimental sum—for the new kind of work in 2014, but the industry was about to explode. As the Instagram user guidelines stated, in a tone as if talking to a child: "When you engage in self-promotional behavior of any kind on Instagram, it makes people who have shared that moment with you feel sad inside. . . . We ask that you keep your interactions on Instagram meaningful and genuine."

"Meaningful and genuine," in this case, just meant that any kind of branding had to look unforced, like it had been an organic decision on the part of the person posting about it. Celebrities were advised to present themselves as relatable and vulnerable, just as advertisers were told to only submit promotions that were aesthetically pleasing, with no price tags.

Instagram employees did want their product to be commercially important, to be big and successful and competitive with Twitter, to contribute meaningfully enough to Facebook that they wouldn't be swallowed up and ruined by the larger company. It was just better if it looked effortless. No journalist would be asked to profile Charles Porch or the community team; instead, the wins were the magazine covers that featured Instagram photos or users.

The company's crowning achievement was an Instagram-related cover on the September 2014 issue of *Vogue*, the most important fashion magazine of the year. It featured Joan Smalls, Cara Delevingne, Karlie Kloss, Arizona Muse, Edie Campbell, Imaan Hammam, Fei Fei Sun, Vanessa Axente, and Andreea Diaconu, with the headline "THE INSTA-GIRLS! Models of the moment in the clothes of the season."

The feature discussed how Instagram popularity was starting to land these women gigs on the most important runways, with the biggest fashion houses, while giving them a voice. Instagram's efforts to visit publications and teach them how to write stories about Instagram were paying off in a big way.

With that, the company finally had the attention of the most powerful person in the fashion industry: Anna Wintour, the editor in chief of *Vogue*. The collaboration was mutually beneficial, she explained. "The girls were using Instagram as a way to introduce themselves to their audiences, and to just talk to audiences in that way through a visual medium, that hadn't been open to anybody before. And for a visually driven title like ours—and indeed, for the company—it just instantly, instantly connected."

• • • • • • • • • • • •

Facebook, meanwhile, was still trying to build a solution to getting more celebrities to use the social network. In 2014, they made an app called Mentions, that celebrities could use to track and communicate with their Facebook fans more easily. They also built an app called Paper that remade Facebook entirely as a more magazine-like experience, similar to Flipboard, putting the focus on high-quality content from publishers. Both products flopped. Besides the inconvenience of being separate apps, requiring a separate download, they were technological solutions to a problem Instagram was solving with human interaction and curation.

Twitter was good at relationships with celebrities and public figures but, unlike Instagram, didn't have any sort of human curation, or an opinion on what ideal Twitter content looked like. Twitter, like Facebook, billed itself as a neutral platform, governed by whatever the masses wanted to see, through their retweeting and commenting on content. Twitter executives would say they were the "free speech wing of the free speech party." It was not their place to get in the way. The biggest missed opportunity was on their Vine app. The crop of homegrown stars there rivaled those on YouTube.

When Vine content production started to slow down, Twitter added a re-Vine button, so people could share other people's Vines in their own feeds. The move had an unexpected side effect, similar to what might have happened to Instagram had they added a re-gram button. Because people could share other users' content in their own feeds, they no longer had a motivation to attempt time-consuming creative skits.

A couple years later, there was little original content production on Vine except from professionals, so those stars realized they had leverage. Twenty of the top Viners banded together to negotiate with Twitter, saying that for about $1 million each, they would post every day for the next six months. If Twitter rejected the deal, they would instead start posting Vines to tell followers to find them on Instagram, YouTube, or Snapchat instead. Twitter refused, the stars abandoned the app, and eventually, Vine shut down entirely.

· · · · · · · · · · · ·

Back in 2014, three months after the *Vogue* cover, Instagram announced that it had reached 300 million users and thus eclipsed Twitter in size. Ev Williams, Twitter's cofounder, finally said publicly what he'd said privately all those times he had passed on buying Instagram: "If you think about the impact Twitter has on the world versus Instagram, it's pretty significant," he told *Fortune*. "Important stuff breaks on Twitter and world leaders have conversations on Twitter. If that's happening, I frankly don't give a shit if Instagram has more people looking at pretty pictures."

What Porch understood, which everyone else eventually would, was that Instagram's power lay not in what was posted there, but in how those posts made people feel. Because there was no re-sharing on Instagram, it wasn't about news and information—it was about individuals, and what they wanted to present to the world, and whether others thought they were interesting or creative or beautiful or valuable. Pretty pictures were just tools on Instagram in the pursuit of being understood and validated by the rest of society, through likes and comments and even money, giving users a small slice of power over their own destiny.

That insight was how Porch won the 2015 Oscars. He thought about it psychologically: what would anyone want, after weeks of workout training to fit into an outfit, after hours of hair and makeup and fittings, with a rare chance to wear an exclusive designer and celebrate a momentous personal accomplishment? Everyone—even the most photographed people in the world—would want a perfect picture.

The company hired Mark Seliger, the noted *Rolling Stone* portrait photographer, and set up a photography studio at the *Vanity Fair* party, complete with Victorian furniture for posing. More than 50 stars, including Oprah Winfrey, Lady Gaga, and *Birdman* director Alejandro González Iñarrítu, posed for Seliger.

All of the resulting portraits were shared, of course, on Instagram— without any corporate fingerprints.

THE PURSUIT OF
THE INSTA-WORTHY

"Facebook buying Instagram was like putting it in a microwave. In a microwave, the food gets hotter faster, but you can easily ruin the dish."

—FORMER INSTAGRAM EXECUTIVE

Instagram was in a luxurious position. Tucked into Facebook, they didn't have to worry as much about the things that other social media companies did. Finding talented employees was easy, as a good portion of the team had worked at Facebook previously and transferred over. New product features could be spun up quickly too because whatever code Facebook built, Instagram could borrow and customize like a template. Facebook's growth team knew all the tricks to help Instagram get to 1 billion users one day. If Instagram wanted to be as big as Facebook, they could copy the strategy.

But Kevin Systrom thought leaning too heavily on Facebook would be dangerous. He did want to be big, but he didn't want to be Facebook. He wanted to recruit the best talent, but didn't want them to bring over

Facebook's grow-at-all-costs values. Instagram, still tiny by comparison, was surrounded by Facebook's culture. Even with more users than Twitter, and almost a third of Facebook's users, Instagram had fewer than 200 employees, compared to more than 3,000 at Twitter and more than 10,000 at Facebook.

Systrom worried deeply about losing what made Instagram special. He wanted the app to be known for its thoughtful design, its simplicity, and its high-quality posts. He focused his team's efforts on preserving the brand, avoiding major changes, and training the app's biggest users and advertisers so that they could serve as models for everyone else.

Unlike Facebook, where employees looked for technical solutions that reached the most users, Instagram solved problems in a way that was intimate, creative, and relationship-based, sometimes even at the individual level if the user was important enough to warrant it. For the Instagram employees, who had such a strong editorial strategy and were always scouting users to highlight, every issue looked like something that could be addressed by promoting the good instead of focusing on the bad. One of their top goals was to "inspire creativity," and so they needed to make sure that the top accounts were indeed inspiring, using the connections built by the partnerships and community teams.

In early 2015, the singer-actress Miley Cyrus, with 22 million followers, was one of those top accounts. That year, she threatened to quit the app, concerned about seeing so much hate and vitriol for LGBT+ youth, especially in photo comments. Instagram found a way to turn her dissatisfaction into an opportunity to land a positive message.

Charles Porch, Instagram's head of partnerships, and Nicky Jackson Colaço, the head of public policy, flew south to visit Cyrus at her mansion in Malibu. They sat around her dining room table, surrounded by art she said she'd purchased off Instagram, and pitched a different plan. She could use the @instagram account as a venue to promote her new Happy Hippie Foundation, which was dedicated to protecting young people who were homeless or vulnerable because of their sexuality or gender identity. Cyrus and @instagram could jointly share thoughtful portraits of people like Leo Sheng, @ileosheng, a trans man, to increase visibility for the people Cyrus was hoping to support.

Cyrus loved the idea and decided to keep using the app, even though Instagram lacked a broad solution to bullying.

Around the same time, the 17-year-old reality star Kylie Jenner was embroiled in controversy because of a viral challenge. The sultry pouts featured in her Instagram selfies were inspiring young girls to try a dangerous body hack: that of putting their lips inside the mouth of a shot glass and then sucking, in order to create enough pressure and swelling to plump them up like Jenner's. Jenner had to reveal that she used temporary cosmetic fillers to achieve the effect, spurring several more news cycles.

In this moment, she remembered that Instagram had told her family that if they ever needed advice on projects, the company could help. So she reached out to see if there was anything she could do to change the public conversation. Liz Perle, the head of teens, had an idea of how Instagram could use the controversy as an opening to promote a more positive message. She sent Jenner a list of ten names of Instagram users who had been vocal about their various body-related concerns. She proposed a campaign where Jenner interviewed these people, then shared their stories on her account, with the hashtag #iammorethan, a sentence that could be completed, as in "#iammorethan my lips."

Jenner wanted to do it, and called to interview everyone on Perle's list herself. The first one she featured was Renee DuShane, a young woman with Pfeiffer syndrome, a genetic disorder affecting her cranial bones. After Jenner shared DuShane's Instagram, @alittlepieceofinsane, with her 21 million followers, both of them immediately received positive media coverage.

● ● ● ● ● ● ● ● ● ● ● ●

Instagram was trying to curate what people talked about and saw on the app so the company would have more control over its destiny. Facebook had proved that the bigger a network became, the bigger the unintended consequences of its decisions. Instagram wanted to borrow what was working without making the same mistakes. Facebook, now with more than 1.4 billion users, had shaped the goals of people and businesses in such a way that everyone was tailoring their content to achieve the top reward on the social network: going viral.

Facebook employees, who were taught that sharing was central to the mission of "connecting the world," applied certain strategies to make it a habit. The algorithm was hyper-personalized, so that any time someone clicked or shared something on Facebook, Facebook would log it as a positive experience and deliver more of the same. But virality had pitfalls. It addicted Facebook's users to low-quality content. The Instagram employees wondered, was a click even an accurate signal of what a user wanted? Or were they being manipulated by the content itself? The viral links had headlines like, "This Man Got in a Fight at a Bar and You'll Never Guess What Happened Next" and "We Saw Pictures of This Child Actress All Grown Up, and WOW!"

Facebook employees had seen their stock options soar in value from rapid growth that came, in part, from not judging their users' choices. They would complain that the Instagram team had the luxury of making different decisions, taking Facebook's resources for granted with an attitude of superiority. Part of that was because Instagram felt it had dodged the virality bullet.

All the editorial work only served to reinforce with Instagram employees that they had succeeded in building a pristine creative paradise on the internet, full of things people didn't know they wanted to see until Instagram showed those things to them. Just like Facebook employees had been indoctrinated into the "connecting the world" mission, Instagram employees were buying their own branding.

But cracks in Instagram's careful, relationship-based plan were starting to show. As more users joined Instagram, the small team became more disconnected from the experience of the average person. For every Cyrus and Jenner, there were millions of others who would never know what it was like to have their concerns heard by an Instagram employee. The ratio was now something like one employee for every 1.5 million users. And Cyrus and Jenner were highlighting real problems, like anonymous bullying and teens striving for perfection, that were systemic, propagated by Instagram's own product decisions, like the ability to post anonymously or compete on follower counts.

Systrom wanted Facebook-size success. He also wanted to avoid cheapening anything about the product, ruining what it stood for. But Instagram was growing so quickly, he couldn't have it both ways. Mark Zuckerberg made that very clear to him—first with the advertising business.

· · · · · · · · · · · ·

Zuckerberg's reality check started in the summer of 2014, about six months after Instagram's first advertisement. Systrom was still reviewing every ad personally, with copies printed out and delivered to his desk. Every big advertiser was trained on how to use popular hashtags like #fromwhereirun and #nofilter, and learn Instagram's tips for a good aesthetic, with a proper focal point and balance in their photos. And it was all going way too slowly for Facebook.

Zuckerberg, who had a year earlier discouraged Instagram from building out its business model, now thought it was time for the app to earn back some of its acquisition price, in the form of revenue for Facebook. Instagram was getting big enough to be useful. Zuckerberg realized that eventually, Facebook's news feed would run out of advertising slots. Nobody had ever grown a network as big as Facebook, and even if they kept adding people, there were a limited number of internet users in the world. By the time the slowdown happened, he wanted Instagram's advertising business to be mature enough to pick up the slack, ensuring revenues kept soaring.

He urged Systrom to increase Instagram's frequency of ads, or the number of advertisers, but mostly to stop being so precious about micromanaging quality. Facebook had its own advertising infrastructure that already made it possible for anyone in the world with a credit card to buy an ad. Just like with the news feed, it was all about personalization; advertisers could say who they wanted to reach, and Facebook would reach that kind of audience for them automatically, with the least human touch possible. All Instagram had to do was plug in and *boom*, they'd have a multibillion-dollar business. Zuckerberg predicted they could do $1 billion in revenue by 2015.

Systrom thought that move, if navigated poorly, could ruin the brand Instagram had built. Sure, they would make money by accepting Facebook's fire hose of ads, but those looked like ads made for Facebook, many of them with tacky text and clickbait wording that would clash drastically with Instagram's aesthetic, and what its users had come to expect in the experience. Facebook hadn't vetted the majority of its advertisers, only their credit cards.

Systrom had some backup. Andrew Bosworth, the ad VP, remembered when he'd had to convince Zuckerberg to ramp up advertising on Facebook, years earlier. He thought Zuckerberg was being a little insensitive to Systrom, given his own reluctance back in the day. He told Zuckerberg that as long as Instagram was selling something totally different than Facebook was, advertisers would be encouraged to take meetings that they could leverage into bigger investments. Plus, wouldn't it be unwise to change Instagram's advertising system in the key planning months before the Christmas season, the most valuable shopping time of the year?

Zuckerberg agreed to wait until January. In the new year, he briefed Facebook's finance team on what he thought each division of the business was going to produce, so they could prepare their 2015 projections for Wall Street. Even though little at Instagram had changed, he told finance he expected $1 billion in revenue from their ads in the next fiscal year.

"Give them six more months," Bosworth argued.

"Six months isn't going to change the situation," Zuckerberg said. "They need to pivot their strategy now."

. . . .

Systrom was called into a meeting with Facebook leadership where he was presented with a chart: Instagram's current ad revenue trend line, juxtaposed with a much steeper line—Zuckerberg's $1 billion goal. If Instagram didn't think it could get there, no problem, he was told. Facebook could help.

Systrom returned to the Instagram office in Building 14 and briefed his team, including Eric Antonow, the head of marketing who had taken over a lot of Emily White's responsibilities. Antonow, an employee at the social network since 2010, was fluent in Facebookese.

"Kevin, you do realize what they're saying, right? They basically told you the number you're committing to," he stressed. Antonow read the political tea leaves. Already, James Quarles, the new head of revenue at Instagram, who had come over from Facebook, was losing battles. Quarles wanted to move thoughtfully with the expansion, with his own sales team that could have a separate relationship with advertisers. He didn't get one. Instead he was able to hire "business development leads" who created training manuals for Facebook's sales force but didn't have ownership over how the conversations went. If Instagram didn't pick up the pace, Facebook might take even more control.

Zuckerberg would eventually force Instagram to open the floodgates and let in ads from any random business buying on the Facebook website. Before they did, for the next few months, the app's engineers raced against the clock to build a system that would save Instagram from death by pixelated digital billboards.

• • • • • • • • • • • •

Instagram was already full of unsanctioned ads—from its own users, paid by businesses to hawk products to their audiences. Instagram employees discussed the idea of trying to take a cut of that market. In February 2015, Twitter paid more than $50 million in cash and stock for Niche, a talent agency that connected advertisers with influencers on Vine, Instagram, and YouTube.

But Instagram ultimately decided they didn't want any part of it. Again, the reason was quality. There was simply no way to personally know all the influencers. If Instagram got involved in the actual transactions, they wouldn't be able to guarantee a good experience for the influencer or the advertiser. Also, as they were building their own ad business, they didn't want to directly encourage a different kind of paid promotion, making the community too overtly transactional.

They tried to focus instead on improving relationships with their users, who were the best reason for more people to join and love Instagram.

Hannah Ray, who previously ran social communities for *The Guardian*, was the first Instagram employee outside of the United States. From the London office, she tried to embody Instagram culture, highlighting its differences from Facebook just as the team in California had. She found an old off-white couch and pulled it to the side of the office. She repurposed some banners from Systrom and Porch's 2013 visit to the National Portrait Gallery to mark the space. Instagrammers were sending postcards from around the world, so she pinned them to the wall. She knew of a local artist who made pillows in the shape of British biscuits and candies, so she commissioned some for the sofa.

Ray was on the community team, so she worked hard to maintain her Insta-shrine amid rows and rows of uniform Facebook desks. There had to always be at least one section of the office that was Instagrammable, she thought.

The project made her quite visible among Facebook sales executives, just in time for some awkward conversations. Often, while she was curating artists or writing handwritten thank-you notes to important photographers, she would be interrupted: *Such-and-such brand wants to do a campaign for a new product. What influencers should they work with to launch it? Could you get us a list of names and contact emails?*

No, Ray would say, *we don't do that.*

But we really should help this important client, the sales team would argue.

We don't want to be the middleman, she'd reply.

Ray would usually appease them by sending over one of the many "Top Instagram Users in X" lists she'd helped the press put together, which were already online and public, and the marketer would reach out to make a deal.

Even that simple action complicated Ray's relationships. The marketplace was small, so the lucky users who won the deals with Facebook clients, many of whom knew Ray, would assume she'd made the pick and thank her. Others, knowing they hadn't gotten a contract, would ask her to give them a chance next time, as they could really use the money.

And so Instagram sometimes ended up being the middleman, unintentionally. Even visits to the innocent Instagrammable office corner could have unexpected economic consequences.

Edward Barnieh, a photographer who was helpful to Instagram because he coordinated InstaMeets in Hong Kong, visited Ray with his wife during a London trip. Ray snapped a picture of the two of them on the office couch with the biscuit pillows, and then they all went off to a pub with a few other photographers. While they were drinking, Barnieh realized Ray had put the photo on the official @instagram account. He had already gained more than 10,000 followers in less than an hour.

After the surge in popularity and the apparent public endorsement from Instagram, the couple soon got their first solicitation from a brand, Barbour, asking if they wanted a free bag to pose with. They agreed. Once Barnieh had done one brand deal, bigger names followed. The idea of an influencer was so new, companies just wanted to pay someone that another advertiser had already trusted. So Barnieh, a Cartoon Network employee, started spending his vacation time on all-expenses-paid trips to photograph parts of Asia on behalf of Nike, Apple, and Sony. He was completely shocked by his luck.

The experience reinforced Ray's fears over her life-changing abilities. "I'm never posting a sofa photo on the main account again," she told Barnieh.

. . . .

As Instagram became more widely used, and as Facebook added pressure to grow and advertise, Instagram employees became at first even more insistent that the app was about beauty and art. The company had just launched five new filters, which it had sent staffers to Morocco to get inspiration in order to build.

It was a frivolous exercise, disconnected from what was happening on the app. Users didn't really care about filters much anymore. Cameras on phones had dramatically improved in the few years since Instagram's launch. And as powerful as Instagram's editorial choices were, they weren't as powerful as the design of the product itself and the incentives it provided to pursue more followers, more recognition, and, increasingly, more money.

By the time Barnieh met Ray, he was already noticing a change in the Instagram community. In Hong Kong, he'd made some of his best friends through InstaMeets and had eventually organized them himself, taking hobbyists on walks where they could share tips and find better angles and lighting. Around 2013, "I was leading locals from Hong Kong in areas they had never been in their own city," he remembers. "It was an intensely positive experience. The intention was truly not to make money or get free things."

But by 2015, some of those enthusiasts had built small photography businesses, making enough money to quit their day jobs. And so Insta-Meets became about business too, because of the opportunity to take pictures with one another. "Someone who was quite outgoing would try to dominate all the pictures at the meet," Barnieh explained. The hustlers' goal was to be tagged in pictures shown to new audiences, possibly increasing their following. An even better prize was to appear on Instagram's suggested user list. "They knew Instagram was watching all these InstaMeets and photo walks and they knew some suggested users would be discovered that way."

Enthusiasts weren't the only ones getting strategic. Barnieh watched new cafes all around the world adopt aesthetics that were popular on Instagram. They would hang bare Edison bulbs, buy succulent planters, make their spaces brighter, fill the walls with greenery or mirrors, and advertise items that were more eye-catching, like colorful fruit juices or avocado toast. In their quest to look modern, he thought they all ended up looking the same, the way airports and corporate offices all look the same. The public was coming to a consensus about what kinds of designs were Instagrammable. Barnieh became more appreciative of his 2013 photos, which now felt like they captured a history before Instagram— and the look Instagram had made popular—was mainstream.

He heard the phrase "Do it for the 'gram" start to catch on. The people who were trying to build businesses off their Instagram photography needed to stand out, so they would venture to picturesque overlooks and beaches, which saw an increase in foot traffic. On the one hand, this quest brought people outside more, and to new locales; on the other hand, it damaged the environment the photos were meant to appreciate with litter and overuse. *National Geographic* wrote about how Instagram was changing travel: visits to Trolltunga, a photogenic cliff in Norway, increased from 500 a year in 2009 to 40,000 a year in 2014. "What photos of this iconic vista don't reveal is the long line of hikers weaving around the rocky terrain each morning, all waiting for their chance to capture their version of the Instagram-famous shot," the magazine wrote.

At one point in 2015, a few Instagrammers in Barnieh's crowd in Hong Kong took the game to another level: they made a habit of hanging off the side of buildings and the tops of bridges. In one shot by Lucian Yock Lam, @yock7, a man is holding another man's arm while he dangles from the side of a skyscraper at night, hovering above a busy street. The caption is a simple hashtag: #followmebro. It got 2,550 likes, a fleeting reward for putting one's life at risk.

· · · · · · · · · · · · ·

Instagram was no longer a niche "community"; it was a mainstream habit. Still, the Instagram employees felt their editorial strategy could make a difference in what users paid attention to. People on the community team decided to be more intentional about who to highlight in campaigns with celebrities, as well as in news articles and on the @instagram account.

They would promote what they felt was their standard fare, like embroidery artists and funny-looking pets. And they would avoid posting anything that perpetuated some of the new unhealthy trends on the app. They would never post a photo of anybody near a cliff, no matter how beautiful, because they knew that gaining a following on Instagram was becoming so desirable that people were risking their lives for perfect shots. They would avoid promoting yoga and fitness accounts, so that they wouldn't seem to approve of a certain body type and make their users feel inadequate—or worse, aroused. They would also avoid promoting accounts that showed off expensive experiences, like ones from travel bloggers.

But sometimes they wrestled with what to celebrate and what to ignore. Should they talk publicly about the #promposal trend, for instance? Teens were developing grand Instagram-worthy gestures to ask each other to the school dance. Was that good for Instagram or perpetuating a pressured culture? And how did they feel about meme accounts? These were wildly popular accounts that weren't about photography at all, but were mostly screenshots of jokes from Tumblr and Twitter. Some Instagram employees were uncomfortable with memes, but also with selfies, bikini shots, and other behaviors that had become mainstream on Instagram, against their artistic sensibilities.

At the very least, they tried to address the fact that the app was becoming a competition for fame. They killed a feature they thought was fueling it: the algorithmic "Popular" page. In its place, Instagram built an "Explore" page, which could be less easily gamed. At first, all of its categories, from food to skateboarding, were curated, handpicked by members of the community team, not via automated selection. There, they chose to embrace some of the new weird corners of Instagram. They had a category called "Oddly Satisfying" that was mostly for videos that were calming and pleasing to watch, like those of people smooshing and stretching homemade slime, carving soap, or slicing kinetic sand.

But the incentives for their users were clearer than ever. Having an audience would always mean having a business opportunity. It wouldn't be long before there were mini-famous slime influencers on Instagram too, going to slime conventions and developing relationships to cross-promote their goo videos.

●　●　●　●　●　●　●　●　●　●　●　●

Liz Perle, the head of teens, thought Instagram should lean into the influencer trend instead of pretending it wasn't happening. In her prior role at *Huffington Post*, she'd been focused on bringing teens to a place where they weren't. Facebook was attempting to lure them with experimental side apps, without much traction. But Instagram had a great opportunity, because it was already full of young people.

She focused on getting to know particular Instagram communities that skewed heavily young, like those for skateboarders and *Minecraft* enthusiasts, and the one centering around #bookstagram, the hashtag for talking about books. She would interview a community's most popular members and then keep track of them on spreadsheets, noting how often they posted, what kind of content they chose, and if they were doing anything unique. If she thought she found a trend, she would urge someone at Instagram or Facebook to help her pull data to see if it was real.

When Instagram launched new features, she tried to make sure they demonstrated them with teen digital-first influencers. The data showed that these kinds of stars, who had become famous on Vine, YouTube, or Instagram, were much more popular than anyone in the office expected them to be. She made a list of 500 of them, then asked Facebook data scientists for help understanding their impact. They found that about a third of Instagram's user base followed at least one of the people on her list.

Perle, like Porch, thought that Instagram should have a role in creating future mainstream celebrities—and that it would be important to build relationships with the ones who hadn't quite become stars yet but had high interest from their audiences. Short of paying them, she could be behind the scenes boosting their careers, giving them a good feeling about continuing to post, maintaining Instagram's relevance among teens.

She suggested teen lifestyle influencer Aidan Alexander, @aidan alexander, be a guest at Arianna Huffington's table at the White House Correspondents' dinner, where he sat alongside Snapchat star DJ Khaled. She let Jordan Doww, @jordandoww, who shared a management company with Alexander, come out as gay on the @instagram account; the public revelation boosted his following by 30,000, after which he was able to quit his day job and just do brand work. When Meghan Camarena, the video game player known as @strawburry17, wanted to host a murder mystery party in Marvel Comics costumes, Perle promised to feature it on the @instagram account and help her attract media coverage.

In return, Perle's teen contacts used Instagram's new products before anyone else, giving feedback and letting the company see what they were creating. And Perle used her insight when sitting in on internal meetings for new product launches, making suggestions about what engineers could tweak to help the product appeal to younger users.

The strategy was a success. Young people were obsessed with Instagram. In 2015, 50 percent of teens in the U.S. were on the app. It became quite important to the structure of their social lives—to the point that it was creating enormous pressure.

The pursuit of followers and influence was a symptom of how aspirational Instagram had managed to make its app. By constantly serving users images of visually appealing lives and hobbies, their community in turn sought to make their lives more worthy of posting about.

Before making a decision about where to go for dinner, tourists would check Instagram to see how delicious their food would look, and so restaurants started to invest more in plating and lighting. Before meeting a new date, users would check out each other's profile to see evidence of interesting hobbies and experiences, as well as of prior relationships. Singles would polish up their feeds. In casting for movies and TV shows, directors would check actors' profiles, to see if they'd bring an Instagram audience if they got a role. Actors needed to become influencers, just as Ashton Kutcher had predicted.

Janelle Bull, a therapist at Anchor Psychology in Silicon Valley, explained that as Instagram became more integrated into everyday life, so too did her patients' anxiety about having an interesting account. Parents worried about giving their children Instagrammable birthday parties and vacations (well before their kids were running their own social media accounts), searching Pinterest or browsing influencer accounts for recipes and ideas that would ultimately photograph well, like special cakes that candy spilled out of when they were cut open. One local parent wanted, for her child's 12th birthday, to rent a party bus to caravan all the kids to Disneyland, so everyone could have plenty of Instagram content. Bull questioned whether children were actually asking their parents to plan such elaborate events.

"Does the parent want attention or the kid?" she wondered. It was becoming a competition. She advised parents that they should take an occasional social media detox to reset their priorities, explaining, "The more you give up who you are to be liked by other people, it's a formula for chipping away at your soul. You become a product of what everyone else wants, and not who you're supposed to be."

She started treating several students at Systrom's alma mater, Stanford University, where the app he created was changing campus life. Now students there agonized about having photos that were compelling enough to get into sororities and fraternities on campus. Networking in those groups would be so important for their future success, they argued. "They worry that without interesting Instagram profiles, they won't get internships, or get noticed by their professors," Bull explained. It wasn't just about their social lives—Instagram had become enmeshed in their professional planning. All around the world, similar stories were playing out.

· · · · · · · · · · · ·

Instagram users invented their own solutions to ease the pressure of gaining likes and followers. Instead of seeking out an Instagrammable life, some sought to invent one. They used photo-editing services to smooth their complexions, whiten their teeth, and slim their figures. They took the idea of the filtered photo one step further, and filtered reality.

On Instagram, it was easy to do. While Facebook was for people using their real identities, Instagram allowed anonymity. Anyone could make an account based on an email address or phone number. So it was quite easy to make more seemingly real people, and sell their attention. If you searched "get Instagram followers" on Google, dozens of small faceless firms offered to make fame and riches more accessible, for a fee. For a few hundred dollars, you could buy thousands of followers, and even dictate exactly what these accounts were supposed to say in your comments. The bot activity sometimes looked suspicious; a fake commenter might post "ur so beautiful!!" below a picture of food.

Purchased fans usually weren't real people, but sometimes gave the purchaser the appearance of having just enough fame to qualify for a brand deal, or to attract more follows from real humans. Fake followers worked on profiles the way Botox works on wrinkles, improving appearances for a few months before Instagram would delete them and reality would return. Instagram learned from Facebook's spam-detection technology to find abnormalities in user behavior. Computers could act in ways a human couldn't, like by commenting hundreds of times in a few minutes.

People rarely admitted to using fake-follower services. Sometimes their denials were true. It was never clear who paid for the fake attention. If not the influencer, was it their talent agency? Someone staffing an ad campaign? The chief marketing officer of the brand itself? Everyone had an incentive to give the impression that their shiny new Instagram strategy was working.

Instagram's detection algorithm was still primitive. When bots looked real, with profile photos and descriptions, following and interacting with real accounts, they were harder to spot. And humans, especially teens, sometimes sent each other messages quickly enough to be mistaken for bots.

The surge in this type of activity couldn't have come at a worse time for Instagram's business endeavors, as the company was in the midst of convincing official advertisers to spend money on the app for the first time. If marketers knew a significant portion of the Instagrammers were bots, they wouldn't be as interested in paying money to reach them. Instagram took its first big swing at the problem in December 2014. Once they thought the technology was ready, they deleted all of the accounts that they thought weren't those of real people, all at once.

Millions of Instagram accounts disappeared. Justin Bieber lost 3.5 million fans, while Kendall and Kylie Jenner lost hundreds of thousands. The 1990s rapper Mase dropped from 1.6 million followers to 100,000, then deleted his account entirely out of apparent embarrassment.

Regular users were affected too because of all the random accounts the bots had followed to appear human. Instagrammers all around the world tweeted furiously at the company, begging for their follow numbers to be restored, saying they hadn't done anything to deserve such punishment. The media dubbed the event "The Rapture."

After all the complaints, Instagram resolved to do its future purges on a rolling basis. And of course the spammers didn't go away; they just got shrewder, working to make their robots look more human, and in some cases paying networks of actual humans to like and comment for clients.

By 2015, dozens of firms, including Instagress and Instazood, offered a compelling service: their clients could focus on perfecting their Instagram posts and they would do all the networking work. Clients would hand over the password credentials for an account, and the services would turn it into a popularity-seeking machine, following and commenting on thousands of others' work in order to be noticed.

For an article he was writing, *Bloomberg Businessweek* reporter Max Chafkin tested out Instagress to see whether it was possible to become an influencer quickly. By the end of one month, he'd spent $10 on automated technology that caused his account to like 28,503 posts and comment 7,171 times, with prewritten generic reactions, including "Wow!" "Pretty awesome," and "This is everything." Those whom he interacted with reciprocated, boosting his audience into the thousands. The project ended when he received his first opportunity for a sponsored post—to model a $59 T-shirt. It's unclear whether the accounts that followed him back were automated too.

· · · · · · · · · · · ·

Instagram discouraged its top accounts from faking it 'til they made it. That kind of growth wasn't sustainable. It was the equivalent of sending users a bunch of notifications to get them to come back to an app. Over time, it would erode trust.

Instagram employees had spent so long comparing their product to Facebook, thinking about how to preserve their artsy, curated idea of Instagram, that the problem of pressure for its users fell to the wayside. They were facing a more immediate demand: the $1 billion revenue goal.

Opening up the advertising floodgates required a delicate political dance. Instagram was determined not to do it the easy way. Ashley Yuki, a product manager in charge of building Instagram's ad system on top of Facebook's, had been a Facebook employee previously, so she knew how to talk to both sides. She had her team sit with the Facebook ad group in a different building so that they could show they were serious about collaborating. Once they'd established a little understanding, an employee on Instagram's team, Hunter Horsley, explained to a Facebook product manager, Fidji Simo, that Instagram needed to keep its ads at least 600 pixels across. That was their minimum.

"Absolutely not," the project manager said. Facebook's limit was 200 pixels across, and if people were buying Facebook ads through the same system, Instagram couldn't have a higher quality requirement. The most important function of the automated system was to *remove friction*, to get rid of any barrier preventing people from spending more money on Facebook.

"What if we increased the pixel requirement on Facebook too?" Horsley asked.

"Then we'd lose a significant portion of advertisers," the PM said.

Maybe. But anyone losing an argument with Facebook had a last resort: running a test to see what the data showed. When Horsley tested whether Facebook would lose money by increasing the quality requirements, he miraculously found that the opposite was true. Advertisers took their ads more seriously, and spent *more*. The change was approved.

● ● ● ● ● ● ● ● ● ● ● ● ●

It seemed like Instagram had won an argument, that growth and quality weren't necessarily at odds. But they needed to compromise on something else. Instagram had only ever allowed square photos. Advertisers were usually shooting in horizontal rectangle formats that could fit in other places around the web, including Facebook.

Instagram's square photos were iconic, so much so that Apple designed a way for iPhones to capture images in that shape. To change the shape of Instagram's photos would be to change Instagram itself, rendering the app unrecognizable, some members of the community team argued. Even though Systrom and Krieger wanted Instagram to make money, they agreed that if the app strayed too far from its roots and capitulated to the needs of the advertising world, it would risk losing everything that made it special.

Yuki, the ad product manager, thought she knew how to get through to them. What if this major problem for advertisers was a problem for Instagram users too? She could see in her Instagram feed that her friends were putting white bars above and below their horizontal photos, or at the sides of their vertical photos, so that they could post the shapes they wanted. She appealed to Krieger, asking if he would at least look into whether it was a common problem. That evening, on the shuttle home to San Francisco, Krieger scanned a sample of two thousand random Instagram photos to test the frequency. The next day he told Yuki that she was right: 20 percent of users were using black or white bars at the sides of their photos.

Because some of the longtime Instagram employees were vocally horrified at the idea of rectangular photos, Yuki was prepared for several rounds of pitching to Systrom. Instead she found him receptive. "I imagine if there was a stadium of people, they would unanimously be saying, 'Why is it so hard for them to make this call?'" Systrom told her. "And that tells me we're holding on to this for the wrong reasons."

When they made it possible for people to post in rectangular shapes, longtime users wrote in, wondering why it had taken the company so long to fill an obvious need.

. . . .

The answer was becoming a little clearer. Distracted by political tussles with Facebook over ads, and meanwhile prioritizing a strategy to appeal to top accounts, Instagram had developed a major blind spot: the experience of the average user. Instagram wasn't thinking enough about those who didn't fit into the curated Instagram brand story.

Just as it had with advertising, Facebook gave Instagram a nudge. The growth team sent Systrom a list of 20 things Facebook wanted the app to change or track so it would add users faster; demands included a more functional Instagram website and more frequent notifications. And they wanted George Lee, a longtime Facebook growth employee, to switch over and run growth at Instagram. Predecessors had failed. A couple years earlier, a few growth employees left in a huff after trying to embed with Instagram, because Systrom was resisting all the ideas he considered spammy. Lee understood that he would be working between two very different cultures.

He told his Facebook growth colleagues, "If I take this job, and I come back and tell you that we're only going to do twelve of them, you've gotta trust me that those are the most important twelve things to do, and that it's coming from me, not Kevin."

And then he had the reverse conversation with Systrom. "I know you got this list of twenty things and not all of them make you super comfortable. But if I tell you that there's twelve things that we actually should do, I need you to trust me."

Systrom argued that Instagram resonated because of its simplicity. He thought that if they changed anything, it would have to be because it made Instagram better, not because it helped Facebook hit a growth metric. Even so, he agreed to hire Lee.

Soon, Instagram's investment in data and analytics would help illuminate something important. It turned out that the high pressure to demonstrate a perfect life on Instagram was actually bad for the product's growth. And it was great for a now-formidable competitor: Snapchat.

THE SNAPCHAT PROBLEM

"What people are experiencing on Instagram is, they don't feel good about themselves. It feels terrible. They have to compete for popularity."

—EVAN SPIEGEL, SNAPCHAT CEO

Facebook's headquarters are optimized for productivity in engineering. The food is free, gourmet, and plentiful, served in themed cafeterias less than a five-minute walk from anyone's office. An employee-only app enables people to peruse menus ahead of time; then there's an option for takeout containers, so people can bring meals back to their desks. For shorter breaks, "micro kitchens" are in every working area, stocked with all manner of healthy and unhealthy packaged snacks, from cereals to wasabi peas and dried mangoes. Coconut drinks and matcha shots chill alongside several different brands and flavors of water, bubbly and flat. When employees are back at their keyboards and done with their snacks, Facebook doesn't dare interrupt whatever work comes next. Each person has a miniature trash can at their feet.

Instagram employees enjoyed these same perks and same waste

disposal privileges until one day in the fall of 2015, when their little bins unceremoniously disappeared. Their personal cardboard boxes, which held miscellany that they had forgotten to unpack as Instagram's space expanded, were diverted to lockers, out of sight. And a few of their giant silver Mylar balloons—in the shape of numbers, to designate an employee's "Faceversary," or anniversary of joining Facebook—had been snipped and discarded.

Systrom told his employees that Instagram was about craft, beauty, and simplicity, and the office needed to reflect this too. He explained that those Faceversary celebration balloons tended to hang above desks long after they had deflated and started to wilt. They should only be tied up a few days at most. The storage boxes made the space look disorganized and unfinished. The trash cans were the worst of all, because they cluttered everything with actual trash. It was time for their space to represent who they were.

In the three years since the acquisition, Systrom had been bothered by the fact that Instagram's headquarters weren't obviously Instagram's. Facebook plasters its walls with motivational posters, printed on-site, with phrases like "Done Is Better than Perfect" and "Move Fast and Break Things," which represent the antithesis of celebrating craft. The previous year, in 2014, Systrom had ripped down some of them in Instagram's micro kitchen, in a rare display of emotion. Then he spent millions on a renovation of the space, especially his own conference room, which he called South Park, after the company's early offices. He adorned it with modern green chairs, a wallpaper patterned with blown-up details of employees' thumbprints, and an acrylic table displaying his first-ever Instagram photo, of his fiancée's sandaled foot plus a dog at a taco stand in Mexico.

Still, he was embarrassed by the place. Systrom had just returned from a management training day at Pixar, where the offices, despite being part of Disney, obviously reflected and celebrated scenes from the animation studio's famous features like *Toy Story* and *The Incredibles*. Kris Jenner had recently phoned Systrom's head of operations, Marne Levine, to ask about visiting Instagram's office with Kim Kardashian. But what was there to visit? The space was pretty much Facebook's still, with the exception of the Gravity Room, a boxy diorama designed specifically for taking pictures, with a table and chairs attached to one side so that whoever stood in it looked like they were walking on a wall. It looked great in social media posts, but in person, it was cracking and peeling, worn from all of the chaotic attention from random visitors of Facebook Inc.

Employees, many of whom had transplanted from Facebook, were not impressed by Systrom's new trash rules. It wasn't practical, and seemed like a distraction from what they should be focused on—their competition. To them, it was a demonstration of peak preciousness, which was a manifestation of Systrom's opinions about the product itself. The idea that the Instagram app was some pristine display of the world's beauty was at best outdated, and at worst a dangerous positioning that could limit its opportunity, ceding market share to Snapchat. A hundred million people were logging into Snapchat every day—a number Facebook could estimate pretty closely from the Onavo tool. Employees had lost confidence that Systrom knew what to prioritize for Instagram's future.

The employees did what twenty-something-year-olds do when uncomfortable: they memed it. They turned Systrom's declaration into a hilarious pseudo-scandal, calling it #trashcangate, or #binghazi, the latter hashtag a nod to the ongoing alarmist news coverage of Hillary Clinton's political fumbles in Benghazi. They would bring it up at Friday question-and-answer meetings with Systrom and Krieger for weeks on end, sometimes just for laughs, because it had become clear Systrom wouldn't budge from his stance. When Systrom went abroad to meet celebrities and sent packages to the office, they would pile up outside his South Park room, where employees would photograph them and snicker at the irony. One employee even dressed up as a trash can for Halloween.

It was harder to talk about the reasons, at the root of the #trashcangate jokes, that they were actually upset. Systrom was focusing too much on what he wanted Instagram to represent, setting a high bar for quality. But Systrom's high bar was exactly what was keeping his team from shipping new features. It was also creating pressure for Instagram's own users, who were intimidated about posting because they thought Instagram warranted perfection.

The real wake-up call about what was wrong with Instagram wasn't coming from Pixar, or from the Kardashians. It was coming from teens.

Third Thursday Teens was a regular evening series, run monthly by research employee Priya Nayak. It allowed Instagram's top management to observe teens in their natural habitat: hanging out on a couch together with their phones. At a nondescript office building in San Francisco, Nayak would sit in a room, facing teens sitting on a couch. Behind her was a mirror, which only reflected on her side. It was actually a window into the room next door, where Instagram product designers and engineers watched the teens, taking in their every word over a bottle of wine.

Instagram's management already had plenty of teen intelligence sourced from Liz Perle's contacts on her spreadsheet of influential young trendsetters around the world. But because these other teens were recruited and paid by a third party called watchLAB and didn't know which company their interview was for, they were more likely to be totally honest about how they felt. Sometimes brutally honest.

The teens revealed that they would meticulously manage their feeds to make a good impression. They had all sorts of unspoken social rules for themselves. They kept track of their follow ratio, and didn't want to follow more people than were following them back. They wanted more than eleven likes on each photo, so the list of names turned into a number. They sent selfies to their friends over group chats to get feedback before determining whether they were good enough for Instagram. They would also meticulously curate. While older users typically kept all their photos up forever, providing a history of every vacation and wedding they'd experienced, some younger people would regularly delete all of their posts, or most, or get entirely new accounts to reinvent themselves as they entered new school years, or wanted to try a new aesthetic theme. If they wanted to be themselves, that was what a "finsta" was for.

Many teens had a separate account called a "finsta"—or "fake

Instagram"—which was actually their more real Instagram, where they could say what they thought and post unedited pictures. But it usually was a private account, shared only with best friends. In other countries, including the UK, teens called them "priv accounts"; in still others, "spam accounts." The names suggested that they didn't want to be judged by what they posted there.

And in late 2015, teens had less need for their finstas, because they could be more real and silly on Snapchat, where everything disappeared shortly after you posted it. Snapchat's Stories feature, especially, was becoming the new way to document their days waking up, walking around school, being bored, hanging out with friends—all activities that might not rise to the level of an Instagram post.

"Instagram," a teen explained one night, "is going to be the next Myspace."

Even though they had been in kindergarten at best during Myspace's glory days, the teens understood the shade they were throwing. "Becoming the next Myspace" was the bogeyman of all tech—the idea that you might be the best thing in the market, until the next best thing catches you off guard and ruins you. In the case of Myspace, the disruptor was Facebook. Paranoia over obsolescence festered at Facebook's very core, and was the reason they'd bought Instagram and attempted to buy Snapchat in the first place.

The anecdotal evidence from Third Thursday Teens was backed up by the data. When Nayak first heard about finstas, she asked Instagram's data scientists to look into how many people had multiple accounts. After weeks of pestering, she got the numbers back. Between 15 and 20 percent of users had multiple accounts, and among teens, that proportion was much higher. She wrote out a report to explain the phenomenon for the Instagram team, since she couldn't find anything about it in a Google search. The team had previously assumed people with multiple accounts had been sharing their phones with family or friends.

To add to worries about users' behavior, the analytics team, led by Mike Develin, unearthed the "reciprocal follower problem." Instagram had emphasized celebrities and influencers too much, and now users' feeds were full of famous people who didn't follow them back. Average people were using the platform just to see what the professionals were up to, and created fewer of their own posts, posting only if a photo met some very high bar of importance or quality. Even then, someone's 14 likes on a post would look paltry alongside Lele Pons's 1.4 million.

Develin's team also found that users weren't posting more than one picture a day. It was considered rude, and even spammy, to take over your followers' feeds by oversharing, to the point that people who did so started using a self-aware hashtag, #doubleinsta.

Instagram was still growing fast. The app reached 400 million monthly users that September, now way past Twitter. But because of the high bar for posting, the rate of posts per user was on the decline. Less content being posted indicated that Instagram was becoming less important in people's lives. It also could mean fewer potential slots in the feed for advertisements. And growth was slowing among Instagram's most important trendsetters: teens in the U.S. and Brazil, who tended to be leading indicators for the rest of the market. Perhaps a more immediate threat than that of becoming the next Myspace was that of becoming the next Facebook: a platform that some teens were not returning to, no matter what the company tried to do to lure them.

All the so-called barriers to sharing were rounded up in a brutal report by one of the company's researchers. The company started on a program to solve them, called Paradigm Shift. To address the finsta trend, Instagram would start to allow people to switch between accounts more easily. For the #doubleinsta problem, Instagram would make it possible to share several photos in the same post. And on and on. Systrom usually wasn't one for war analogies, but if they were at war with Snapchat, Paradigm Shift was their beachhead, he would say.

And yet, in the opinion of a small minority of Instagram employees, Paradigm Shift was more evolution than revolution, unlikely to change the underlying trends. While Systrom was finally willing to make changes, these employees thought they weren't drastic enough, and that Instagram had to do something bigger and bolder. They absolutely had to introduce some kind of way to post things that disappeared, à la Snapchat Stories, to reduce the pressure to be perfect on Instagram.

Nobody wanted to hear about it, least of all Systrom.

• • • • • • • • • • • • •

As far as Systrom was concerned, high standards allowed Instagram to thrive. Systrom was the ultimate self-improver. In the last few years, besides building a social network with 400 million users, he'd gotten better at searing steaks, at running miles, at understanding interior design, at raising a puppy. He was working with an executive coach. And he was ready for a new challenge, which would allow him to get in shape and appreciate the natural beauty of the San Francisco Bay Area.

The Bay Area is full of cyclists, risking their lives zipping around the blind turns of the hills that overlook the water, in their branded, padded spandex shorts and neon zip-ups. But most of the riders are not pros, just (mainly) men who take their hobby quite seriously. Cycling is a popular way of unwinding from the always-on stress of the technology industry, allowing one time to think. Systrom had absorbed enough of this religion by osmosis, and in late 2015, he found a place to start: the Bay Area's cycling mecca, Above Category.

The cycling shop, north of San Francisco and a couple blocks from the Sausalito marina, was globally recognized for its rare selection of high-end gear, including bikes worth tens of thousands of dollars. Systrom could have afforded any of it, but first he wanted to *earn* it. He told Nate King, a tall brunette with floppy curls who was working that day, that he wanted a bike that was not too flashy, just to help him get started.

King fitted him and helped him order a custom Mosaic road bike. Systrom put it on a stationary mount in his San Francisco home. Every morning, Systrom would pedal, running through in his mind all of the things that needed to be done. He and Nicole Schuetz were getting married on Halloween, with a black tie masquerade ball in the Napa wine caves; celebrity designer and friend Ken Fulk would bring the couple's vision to life with Victorian flair, and *Vogue* would feature the event. They were planning a honeymoon in France. Instagram had been able to reach $1 billion in revenue run rate at record speed, in just about 18 months since those first ads ran, thanks to Zuckerberg's pushing. So much had changed, so quickly.

As Instagram got larger, he agreed with Facebook that he needed to think more often about data and start measuring Instagram the way he measured his coffee extractions and ski runs. Based on the information, they could tweak the strategy slightly, until the numbers were better. That was what Paradigm Shift was about. It was a Facebooky approach that at first had seemed antithetical to Instagram's intuitive design culture, but would be valuable.

He measured his cycling progress on the multiplayer game *Zwift*, becoming obsessed with topping his personal best. Nate from the shop became his cycling mentor, and a frequent recipient of Systrom's random emails asking how to enhance his strategy: *Do I need a power meter? How about clutches?* Eventually King would take Systrom out on some more challenging rides, with his more serious bikers from the industry. Systrom protested at first with a little self-deprecating humor, saying, "I'm not good enough."

"Dude, you created a verb!" King responded, and that was enough to get Systrom going.

That verb—"to Instagram"—was another thing for Systrom to ponder on his rides. To him, it meant capturing the highlights of life, what was important, beautiful, or creative. But Systrom's experience was unique. Because of his job, he was surrounded by a frequent assortment of beautiful and interesting things. It would not be a stretch to say he had one of the most beautiful, interesting lives of all Instagram users.

In July, he was boating in Lake Tahoe, where he owned a lakeside cabin decorated by Fulk. In August, he was vacationing in Il Riccio off the coast of Italy, then night diving in Positano. In September, he got to dine with Kendall Jenner and designer Olivier Rousteing during Paris Fashion Week. In October, he met the president of France, François Hollande, and helped him join the app. A few days later, he took selfies with actress Lena Dunham and photographer Annie Leibovitz. And that was just a sampling of what he posted publicly; he didn't reveal, for example, that he got to meet Hollande's dog and sample fancy chocolates in the wine cellar of the Élysée Palace.

Systrom, like the teens, was posting less frequently on his own feed, putting only the best stuff, curating and deleting things he didn't want to be part of his permanent record. Plus, he had 1 million followers now, and needed to represent the company. It wasn't like the early days, when users were going on photo walks to find beauty in unexpected places.

"Instagram is not for half-eaten sandwiches," he would tell employees, setting up a contrast to Snapchat's rawness. On a scale of quality images, rated 1 to 10, Instagram was for those ranked 7 and above, Systrom would say. If they changed that, they might ruin it. The plan might have been called Paradigm Shift, but the philosophy was still "don't fuck it up."

* * * * * * * * * * * * *

Employees had been going around Systrom to take risks. Earlier that year, team members had an idea for a feature called Boomerang that would allow people to take a quick succession of images that would combine into a short video, playing forward and then reversing, and then forward, and then reversing. It made simple movements entertaining: cake would be cut and uncut, water would be spilled and unspilled, over and over. Instagram employees John Barnett and Alex Li, expecting the idea would be rejected by Systrom, didn't approach him about it. Instead, they built Boomerang at a Facebook-sponsored hackathon, and it won. Systrom then felt confident enough to release Boomerang as part of Instagram, after which he received a congratulatory email from Zuckerberg.

Barnett and Li had spent many afternoons at the on-campus Philz— the only place at Facebook where you had to pay for coffee—scheming about how to convince Systrom that Instagram needed a way to post things that disappeared. They were both in the Paradigm Shift group, but any time they had a serious discussion about a Stories-like feature, it spurred drama.

Li was getting especially anxious. His wife was due to have their first child in a couple months, around Thanksgiving, and if he didn't do *something* to fix Instagram before going on parental leave, he was going to be frustrated his entire time away from the office.

Eventually, he decided he needed to cut through the layers of management between him and Systrom so he could make the pitch directly. Li explained to Krieger what he was thinking. *Put me in, coach,* he begged. Krieger wasn't the decider, but he was still a founder, as well as a sympathetic ear. He was always good at listening and diffusing conflict. Krieger agreed that something like Snapchat's Stories was worth thinking about, but said he wasn't going to advocate on Li's behalf.

One evening, Krieger got tired of Li's lobbying and relented. "We should just get on a call with Systrom right now," he said. "He's probably driving in his car."

Systrom answered, and Li launched into the impassioned appeal he'd been waiting so long to deliver. He explained that between him, Will Bailey, and John Barnett, there were good people who cared so much about making this happen that they would build it on their free time.

"I'm tired of hearing this shit," Systrom said. There was already a plan in place. They needed to agree to disagree.

After the tense phone call, Li was so amped that he spent the rest of the night at the gym, shooting basketballs. Then he wrote a long email to Systrom asking for a compromise. Could they at least have a regular smaller meeting, with him, Barnett, and Bailey, where new ideas could be more thoroughly discussed? Systrom told him to be patient.

● ● ● ● ● ● ● ● ● ● ● ●

In the fall of 2015, Ira Glass hosted an episode of *This American Life* on National Public Radio called "Status Update." It opened with three girls, 13 and 14 years old, explaining how Instagram was putting pressure on their entire social lives. The teens, named Julia, Jane, and Ella, explained that in their high school, if they didn't comment on one of their friends' selfies within ten minutes, those friends would question the entire nature of their budding relationship.

In their comments, they used super-affirming language: "OMG you're a MODEL!" or "I hate you, you're so beautiful!" Often it was accompanied by the heart-eye emoji. If the selfie poster cared about the friendship, they would have to comment back, within minutes again, with a reply like "No YOU'RE the model!" (Never "thank you," which would imply that they agreed they were beautiful, which would be horrifying.) The girls expected 130 to 150 likes on their selfies, and 30 to 50 comments.

The conversations on Instagram—especially the nature of who was commenting on whose photos, and who showed up in whose selfies— were what defined their friendships, their social standing at high school, and their personal brand, which they were already acutely aware of. As they explained to Glass on the radio show,

Julia: To stay relevant, you have to—
Jane: You have to work hard.
Ella: "Relevance" is a big term right now.
Ira: Are you guys relevant?
Ella: Um, I'm so relevant.
Jane: In middle school. In middle school, we were definitely really relevant.
Ella: We were so relevant.
Jane: Because everything was established. But now, in the beginning of high school, you can't really tell who's relevant.
Ira: Yeah. And what does relevant mean?
Jane: Relevant means that people care about what you're posting on Instagram.

Glass explained in his narration that it was because of this pressure that stakes were so high. They limited themselves to only the best selfies, which were carefully approved ahead of time in group messages with their girlfriends. It was in those same chats that they would screenshot and analyze other kids' bad selfies and comments from their school.

"Each of them only post a couple pictures a week," Glass explained. "Not that much of their time on Instagram is being told they're pretty. Most of it is this, dissecting and calibrating the minutiae of the social diagram."

• • • • • • • • • • • •

The episode was passed around heavily at Instagram headquarters. This was the exact kind of behavior that Li and Barnett were concerned about.

Barnett, a gentle, bearded product manager, had been emboldened by his managers saying in a recent performance review that he was too nice, that he should be more of an asshole about his ideas. But he too would get shut down after raising his hand in the Paradigm Shift meetings to pitch a version of Stories. His managers told him not to push it and to stop talking to his colleagues who were interested in building it, because Systrom had clearly made up his mind.

By January, the stress of the battle had worn him down. In a meeting with Systrom, while sweating profusely, Barnett mustered up as much assholery as he possibly could, telling the CEO that the current Paradigm Shift plan was not effective or inventive enough to beat Snapchat.

Systrom was unmoved. "We will not ever have Stories," he said. "We shouldn't—we can't—and it doesn't fit with the way people think and share on Instagram."

Snapchat was a totally different thing, and Instagram could come up with its own ideas.

Defeated, Barnett made a plan to transfer to a different part of Facebook. But not before he convinced some employees to secretly work on a mock-up, hidden away from Systrom in Building 16. Christine Choi, who had helped design Boomerang, worked with him to create a concept for displaying content that disappeared after 24 hours, arranged in little orange circles at the top of the app. She uploaded it to Pixel Cloud, the internal design-sharing system. Barnett was advised not to show it to Systrom.

• • • • • • • • • • • •

Systrom had good reasons to avoid taking the plunge into Stories-like tools. All of Facebook's copycat attempts had failed, starting with Poke, the blatant remake of Snapchat that had failed so badly it convinced Zuckerberg to make his $3 billion acquisition offer in 2013. Afterward, when Facebook spun up their internal Creative Labs Skunk Works to make apps that would appeal to teens, all were short-lived. There was the Slingshot app for photo responses to ephemeral messages. There was also an app called Riff, a take on Snapchat Stories, which was barely significant enough to be mentioned in the media. None of them garnered more than a few thousand users.

Mark Zuckerberg himself had explained, in an internal memo to executives that winter, that the tools related to the phone camera would be at the core of Facebook's future. He suggested that some form of ephemeral sharing was going to be on the Facebook road map, and that perhaps Instagram should consider it too. But fast-following, as it was called in the technology industry, rarely worked.

"Rivalry causes us to over-emphasize old opportunities and slavishly copy what has worked in the past," venture capitalist and Facebook board member Peter Thiel wrote in his 2014 book *Zero to One*, which Systrom asked all his managers to read. "Competition can make people hallucinate opportunities where none exist."

Systrom was starting to get deep into another book, by former Procter & Gamble CEO A. G. Lafley, called *Playing to Win*. Lafley's theme resonated with the Instagram founders' focus on simplicity. "No company can be all things to all people and still win," Lafley wrote. First companies had to pick where to play; then they had to decide how to win in that market, without worrying about everything else.

Incidentally, Lafley had just started mentoring Snapchat CEO Evan Spiegel. And Spiegel had decided where he wanted to play: Instagram's turf.

· · · · · · · · · · · · ·

Systrom might have been the only Silicon Valley executive with a somewhat viable excuse to attend the Academy Awards. He wanted to see and be seen by some of Instagram's most high-profile users, to understand how they were sharing on the app. In 2016, he put on his tux and brought his sister Kate as his plus-one, posting a black-and-white mirror selfie of the two of them together on his Instagram before heading out to the red carpet.

While Systrom was mingling, stars were posting on Instagram more than they ever had. But as he looked at what they were saying, he noticed a trend. A lot of them were using their posts to refer fans to more exclusive behind-the-scenes videos—on Snapchat.

Krieger had noticed the same thing when he attended the Golden Globes earlier that year. Instagram had taught all of these people the value of communicating directly with their audiences, without a publicist or the paparazzi. But Instagram wasn't allowing them to share as much as they wanted to, just because of the way they had built the app. It turned out stars had the same trouble teens did: they didn't want to overload their followers or post things that would last forever.

The media also picked up on the trend. "While we adored the many Instagram and Twitter pictures posted during the big night, several of our favorite A-list celebrities added a new social media outlet to their Oscars extravaganza: Snapchat," *E! News* wrote. Kate Hudson played with Snapchat's silly face filters and took selfies with Hilary Swank. Nick Jonas snapped himself hanging out with Demi Lovato at the *Vanity Fair* party. Most intimate of all was Lady Gaga, who brought her Snapchat viewers with her as she was getting her makeup done pre-show. She then revealed how nervous she was about performing "Til It Happens to You" with onstage guests who were survivors of sexual abuse.

Snapchat had made it easier for sites like *E!* to cover the event by making it possible to view stories on the web, not just mobile phones, for the first time. It seemed Snapchat wasn't just for "half-eaten sandwiches," as Systrom had dismissed it; it was a way to give every person their own reality television show.

Krieger and Systrom realized that this was what Li, Barnett, and others had been trying to tell them: Instagram users now had a place to put all the content they would otherwise leave on the cutting room floor. If they didn't make it possible to put that content on Instagram, they might lose those people forever to Snapchat.

You're at a fork in the road, Systrom thought to himself. *You can either stay the same because you want to hold on to your idea of Instagram, or you can bet the house.*

He decided to bet the house. Systrom was fully aware that if he failed, he could be fired, or ruin everything. But at that point, the only failure that could be certain was if he decided to do nothing.

The exception was one Thiel had written about in *Zero to One*: "Sometimes you do have to fight. Where that's true, you should fight and win. There is no middle ground: either don't throw any punches, or strike hard and end it quickly."

The need to move quickly wasn't just about Snapchat. If some kind of ephemeral sharing was going to be on Facebook's road map, Instagram needed to build their attempt first. Otherwise, it would lose its cool factor.

• • • • • • • • • • • •

Soon after, Systrom arranged an emergency meeting for all his top product executives. On a whiteboard at the front of the South Park conference room, he drew a mock-up of the Instagram app with little circles at the top of the screen, and passed out a document with Choi and Barnett's concept—which simultaneously shocked and flattered them. He explained that every user would get to add videos, which would disappear within 24 hours, to their personal reel and that he wanted the team to launch this new feature by the end of the summer. To most people in the room, it felt dramatic and novel, a moment where they were inspired by their leader, who was finally willing to take major risks. "It was like being in the room when John F. Kennedy announces you're going to the moon," one executive later recalled. Few people knew the tension behind the decision.

Systrom and Krieger felt especially confident that they would be able to ensure this new project wouldn't just be a straight copycat, but a thoughtful product exercise, because they had hired some people they trusted to get it done.

Robby Stein, for example, was Systrom's coworker from Google long ago, who had sent him a congratulatory email when Instagram first launched. Now, lured by Systrom's willingness to make dramatic changes, he would join the team and help specifically with thinking about how friends talked to each other on the app.

And then there was Kevin Weil, a friend of Systrom's and fellow exercise enthusiast, who was Twitter's head of product, working under then CEO Jack Dorsey. Twitter now considered Instagram, not Facebook, to be enemy number one, especially given all the work Instagram had done to get public figures to use the app. But the company was recovering from a streak of layoffs and executive departures, including Dorsey replacing Dick Costolo as CEO. Dorsey was having trouble making major product decisions to reverse Twitter's slowing growth. Weil needed to get out of there. He interviewed for several different kinds of jobs, including at Snapchat, where Spiegel was so confident he'd join that he introduced Weil to his most trusted employees, on the secretive design team.

The news that Weil was leaving Twitter to become Instagram's head of product broke during an executive off-site at the end of January, where Twitter was planning its goals for the year. Dorsey was blindsided and visibly upset. While he'd known Weil was leaving, he'd been under the impression it was to take a break, not go to a major competitor. Weil was escorted off the premises, and then Dorsey wrote an angry email to Twitter's entire staff about his disloyalty.

When Weil arrived at Facebook headquarters, he had just received both texts and direct Twitter messages from Twitter's head of revenue, Adam Bain, marking the end of their friendship. Weil was shaking, wondering if he had acted unethically. Sheryl Sandberg called him into her office to calm him down.

"We're media companies, in the same line of work," Sandberg explained. "Imagine if you worked for ABC or CBS, and then got recruited by NBC. Would it be unethical to go there?"

Weil supposed not.

Jack eventually apologized to Weil for his anger, which had deep roots in his own feeling of betrayal after Instagram's sale to Facebook so many years earlier. Spiegel, always paranoid, decided Weil had probably been spying on behalf of his new employer, and put a moratorium on hiring anyone from Instagram for about six months. The only thing left for Weil to do was prove that he'd made the right career decision.

• • • • • • • • • • • •

Through Charles Porch's strategy, Instagram was getting closer to unseating Twitter as the number one destination for pop culture on the internet. But Twitter still had something Instagram didn't: the pope.

A month after deciding to lower the pressure on Instagram users through optional disappearing posts, Porch and Systrom were still going strong in their efforts to sign up famous people. Anna Wintour, *Vogue*'s editor in chief, agreed to host a dinner for Systrom and big-name designers during Milan Fashion Week, as she'd previously done for him in London and Paris. Guests included Miuccia Prada, Silvia Venturini Fendi and her daughter Delfina Delettrez-Fendi, and Alessandro Michele, the creative director of Gucci.

If they were going to Italy anyway, Porch thought, they might as well dream big. They scheduled a meeting with the prime minister, and then thought, *Why not try for the pope?*

Facebook had contacts with the Vatican, which Porch leveraged to request a papal audience for Systrom. He had a strategy argument. The Catholic Church, with a network of 1.2 billion, smaller than Facebook's, needed to stay relevant. It could use Instagram to reach a young audience. Miraculously, Pope Francis agreed to meet, just two years into his papacy.

It's customary to give the pope a gift, so Instagram's community team put together a light blue hardcover book of images on the app that would speak to issues important to Pope Francis, such as the refugee crisis and environmental preservation. After Porch and Systrom arrived at the Vatican and had a pre-meeting with Italian priests, Swiss guards escorted Systrom into a private meeting with the pontiff. There, he had a few minutes to make his case directly.

Pope Francis listened intently, then said he would consult with his team about the idea of joining Instagram. But ultimately it was not up to them. "Even I have a boss," he said. He gestured toward the sky.

A few weeks later, Porch got a call. Pope Francis would make an Instagram account. He wanted Systrom to be at the Vatican for the occasion, in about 36 hours. They jetted over.

The whole Vatican press corps was around to film and report on the occasion. And everything was set: the handle, @franciscus, and the first photo, a profile shot of the pope kneeling on a red velvet and dark wood prie-dieu, eyes closed and head tilted in solemn reflection, in ivory mozzetta robes and zucchetto skullcap. The pope's first post was a call to action: "Pray for me," he wrote. With one tap on the papal iPad, it was live.

The pope's new account became international news, with that first post in March 2016 garnering more than 300,000 likes. The moment was a crowning achievement of Instagram's strategy to get the most significant people in the world to use the app, initiated by Porch, with his celebrity wish list, supported by Systrom's frequent jet-setting and strategic schmoozing over wine and Michelin-star dinners.

Systrom spent that night indulging in one of his favorite Roman dishes: pizza, which of course was from a spot he had extensively researched. What he didn't let on to anyone at the time was that he wouldn't be doing many more of these trips.

Instagram had been too focused on its biggest users. It was time to think about everybody else.

• • • • • • • • • • • • •

All of the polished activity from high-powered accounts, over time, would mean little without a base of regular people coming back to the app every day to see what their friends were up to. With that same reasoning in mind, the founders made a separate major decision. It was not as controversial inside the company as it ended up being outside.

Up until this point, all of the content on Instagram had been arranged with the newest posts first. But Instagram's chronological feed had become problematic and unsustainable in terms of keeping everyday people engaged. The more professional Instagrammers tended to post at least once a day, at the most strategically viable time, with content they expected would get the most likes, while more casual users might post less than once a week. That meant that anyone who followed a combination of influencers, businesses, and friends would log on and then most likely see content from professionals at the top of their feed, not the posts from their friends. It was bad for their friends, because they didn't get the likes and comments they needed to be motivated to post more, and it was bad for Instagram, because if people didn't see enough amateur posts, they were more likely to feel their own photos were unworthy by comparison.

Their best solution was an algorithm that would change the order of the feed. Instead of putting the most recent posts at the top, it would prioritize content from friends and family over that from public figures.

They decided the algorithm wouldn't be formulated like the Facebook news feed, which had a goal of getting people to spend more time on Facebook. Instagram's founders reasoned that "time spent" was actually the wrong metric to aim for, because they knew where that road had led Facebook. Facebook had evolved into a mire of clickbait video content produced by professionals, whose presence exacerbated the problem of making regular people feel like they didn't need to post.

Instead Instagram trained the program to optimize for "number of posts made." The new Instagram algorithm would show people whatever posts would inspire them to create more posts.

Instagram did not explain this publicly. They essentially told the public, *Your feed will be better. Trust us.* "On average, people miss about 70 percent of the posts in their Instagram feed," Systrom said in the company's announcement. "What this is about is making sure that the 30 percent you see is the best 30 percent possible."

But people mistrusted algorithms, in part because of Facebook. To users of Instagram, the change felt like an affront to the experience each of them had worked so hard to curate and control. The launch drew immediate backlash. When Instagram ran blind tests, users liked the algorithmic version more; when told it was algorithmic, they said they preferred the chronological version.

While regular users got more likes and comments, the most prolific users saw a dramatic slowdown or stop in their growth. Influencers and brands had built growth into their business plans, and now, with this algorithm, it was gone. Instagram had an unsatisfying solution for them: they could pay for ads.

Systrom told his team they needed to have conviction that the algorithmic version was, in fact, better for most people. By then, Instagram had 300 million people using it daily, triple Snapchat's user number. With the idea of reaching Facebook's size in the realm of possibility, Systrom put it in perspective. "If we're going to get to a billion, that means seven hundred million people are going to join Instagram who have never experienced a ranked feed," he said, sounding more like Zuckerberg than he ever had. "You have to care about the community you have, but you also need to think about the people who have not even experienced the product, and don't have any preconceptions."

Still, the public's bitter opinions around the algorithm explain why, when Instagram's engineers were developing the disappearing-stories tool, they had no idea whether it would be well received.

. . . .

This public outcry over the feed intensified the debates the Instagram team were having over Stories, stressing about every tiny detail. People would only use the product if it felt right, so what made sense? Should Instagram allow users to upload their phone's camera roll content into Stories or make them use the camera within the app? Should Instagram let people build a separate friend network for Stories or automatically allow them to share Stories to all their friends? Should the bubbles at the top have pictures of people's faces, or pictures of the content they were producing? Eventually, when it came time to add advertising to the experience (because it was Facebook, so there was going to be advertising), should those brands get to have bubbles too?

"Reels" was the code name at Instagram, but everyone was casually calling the product Stories. In a computer-filled conference room called Sharks at Work, with a glass garage door, Systrom and the others would spend hours drawing out various possible versions on whiteboards. They were mostly trying to decide what the simplest solution was. For example, Instagram didn't need to launch with the tools Snapchat had, like face masks that used image technology to let people digitally wear cartoon puppy ears or barf rainbows. They reasoned they could add that kind of thing later.

Will Bailey and Nathan Sharp, the engineer and product manager leading Stories, spent so many hours in the office during this rush period that their team often spent the night rather than endure the hour-long drives to and from San Francisco. Barnett saw one of the engineers post to the test version of Stories in the middle of the night, with tears and eye-bags drawn on their selfie, and alerted his former Instagram colleagues— couldn't someone help them out? At first, their managers provided them with Instagram-branded pillows and blankets. Eventually, they were allowed to expense stays in nice local hotel rooms.

Meanwhile, the head of research, Andy Warr, tested the product with anonymous outsiders sourced by watchLAB. As he interviewed research subjects, Systrom and the others watched how people interacted with the app from behind the one-way mirror.

"Which company do you think made this?" Warr would ask the research subjects.

"Probably Snapchat," they responded.

· · · · · · · · · · · ·

In all of its Snapchat copycatting, Facebook was forced to learn, over and over, that just because it had made one world-changing product didn't mean it could succeed with another, even when that product was a replica of something already popular. Snapchat, meanwhile, learned that it could ignore Facebook's repeated attacks. In fact, Facebook was so apparently unthreatening during this period that a Snapchat executive proposed trying something crazy: being friends.

Snapchat's best asset and biggest problem was Evan Spiegel himself. Success had gone to his head, and now he was building a company based primarily on his personal taste, not according to any sort of systematic decision-making. His employees saw him as stubborn, narcissistic, spoiled, and impulsive. Spiegel hated product testing, product managers, and optimizing for the data—basically everything that had made Facebook successful. The result was a company full of yes-men (and a few yes-women) hanging on Spiegel's every word, who expected they would be fired if they disagreed with his direction. His executives' tenures tended to be short; Emily White, Systrom's early business helper who went to be Spiegel's chief operating officer, lasted just over a year.

Spiegel needed a mentor who would help him grow up. Imran Khan, his chief strategy officer, reasoned that there were only two people in the world who might have the ability to get through to someone who'd dropped out of school and become really rich really fast: Mark Zuckerberg and Bill Gates, both of whom had the same lived experience.

Cozying up to Zuckerberg was tricky, strategically, because he was still holding a serious grudge against Spiegel for emails leaked to *Forbes* discussing the $3 billion acquisition attempt in 2013. Worse, Spiegel still felt pretty strongly that Facebook was inherently evil and uncreative. Khan decided to start with his Facebook counterpart, Sheryl Sandberg. He reached out asking if it was possible to repair the relationship, and she agreed to meet at Facebook's headquarters.

In the summer of 2016, Khan made the trip from LA to Menlo Park. Sandberg had made some arrangements up front to keep his visit confidential. He took a secret entrance, avoiding the general security check-in, so employees wouldn't recognize him and get the wrong idea. That was perhaps the first sign that Facebook had a different agenda than he did.

Sandberg had invited Dan Rose, Facebook's partnerships head, to join the conversation. She started out with a little friendly condescension, explaining how very difficult it was to build a major advertising business. She would really love to be a resource for Snapchat in any way she could, she said. Khan humored her until midway through the meeting, when she excused herself. Then it was just Khan and Rose.

"There actually is a way we could help," Rose said. "We could buy Snapchat." He explained that the company would end up just like

Instagram—totally independent, but applying everything Facebook had learned to help the business scale more quickly.

There's no way Spiegel would go for that, Khan thought, but they did need the money. They were severely unprofitable after spending so much on data storage with Google. "How about a strategic investment?" he asked.

"We don't do that," Rose said. "We buy, or we compete."

• • • • • • • • • • • •

Meanwhile Instagram, oblivious to these conversations, was intent on striking Snapchat hard and ending it quickly.

Facebook usually launched something to a tiny percentage of its user base, around 1 or 2 percent, to see how people reacted. Then it could roll the new product out to 5 percent, or a couple countries, before eventually reaching the rest of the world. Zuckerberg thought it was important to gather data on how a product would affect the company's underlying usage metrics. Facebook also tended to release products half-baked, and use the feedback to tweak them in real time.

The Instagram team was going to try the opposite: launching Stories, at least a simple version of it, to all 500 million of its users at once. They called it a "YOLO launch," after the acronym for "you only live once." It was an extremely risky strategy by Facebook standards, but Systrom couldn't be convinced otherwise. He thought it was such a big change that everyone needed to be able to access it, or else it would be starved of the oxygen it needed to work.

Robby Stein, the product director in charge of Stories, would later compare the anxiety around the launch to that of a major life event, like getting married or having a child, where you have convinced yourself it's a good thing and anticipated it for many months, but you know everything will be forever changed once you do it.

For Zuckerberg, it was also a last chance. A couple months after Khan's meeting with Sandberg, and a few days before Instagram was set to launch Stories, the Facebook CEO called Spiegel on his cell phone. "I heard you've been talking to Google," Zuckerberg said. "Facebook would definitely be a better fit." If anything, Zuckerberg said, Facebook could make an offer so rich that Google would have to go higher.

Spiegel played it cool. "We're actually not talking to Google," he said. "But if we ever do, I'll let you know."

The door for a sale was officially back open. And Systrom, the poster child for a successful Facebook acquisition, who had been instrumental in getting WhatsApp to sell, was totally in the dark about Zuckerberg's conversations with his biggest competitor. So was Snapchat's board. Spiegel never told them about the call, because, just like at Facebook, Spiegel and his cofounder held the majority of the voting control, rendering everyone else's opinion irrelevant.

· · · · · · · · · · · ·

On the day of the launch of Stories in August 2016, the whole team arrived around 5 a.m. at Facebook's headquarters, which were otherwise empty that early. In the Sharks at Work conference room, they stood around with breakfast burritos, which had been catered because none of the cafeterias were open yet. Supporters showed up until it was standing room only around Nathan Sharp's computer.

"FIVE, FOUR, THREE, TWO, ONE," the team counted down, and Sharp pushed a button to send Stories live to the world at 6 a.m. PST. Everyone watched as the numbers climbed. A couple employees snuck some celebratory bourbon into their coffee when Systrom wasn't looking. The office now had a glass case full of expensive bottles.

Barnett, who was now working on Facebook's youth team, came to see what he'd advocated for finally come to life. Systrom came up to him to congratulate him. "Sorry I unfollowed you on Instagram," he said. Barnett had been posting too much. "I'm going to follow you again right now."

· · · · · · · · · · · ·

Systrom had told his communications team that he would acknowledge to the press that the Stories format was a Snapchat invention that Instagram had copied, and that was why they had the same name. (*"You're going to do WHAT?"* Facebook PR head Caryn Marooney exclaimed. Usually

Facebook would spin any copied products as a "natural evolution" of what users wanted.)

It was a good instinct because that was how the press evaluated the move anyway. All the major headlines used some version of the word "copy" in them. By not denying it, Systrom took the momentum out of the criticism. He explained that it was just a new form of communication, like email or text messaging, and that just because Snapchat invented it didn't mean that other companies should avoid using the same opportunity.

He held an all-hands meeting for the Instagram staff, explaining how Instagram's Stories managed to be innovative despite the competitive inspiration. Plus, the tension over how to solve the problem had helped everyone deliver a more polished result. Employees came up to him after, thanking him for the inspirational talk.

Though plenty of users complained about Stories on social media, the numbers showed that they were indeed using it, more and more every day. It took a while to catch on in markets that Snapchat dominated, like the U.S. and Europe, but immediately took off in Brazil and India, where Snapchat's product kept breaking with weaker connections on Android phones. Instagram had launched it with perfect timing, right before teens returned to school.

Andrew Owen, on the community team, had spent the previous few months trying to get important users to start posting video on Instagram, focusing on action-packed events like the X Games. He kept getting rebuffed; everyone wanted to use Snapchat instead. But when Instagram Stories launched, he was in Rio de Janeiro with Justin Timberlake, who was performing at the Summer Olympics. Backstage, Timberlake was hours early for his performance and bored as Owen pulled up the Stories option on the @instagram account. Timberlake took the phone and started filming as he chatted with fellow performer Alicia Keys, creating content for all the millions of followers of @instagram. The next day, Owen did the same thing for the @instagram account with the U.S. women's gymnastics team.

The community team was responsible for filling the corporate account's stories with interesting content every day. That way, everyone

following it would always have something to watch, helping them understand how to use the new product. Community team member Pamela
Chen traveled to New York to teach Lady Gaga about Stories since the
singer was promoting a new album. After Rio, Owen went to Los Angeles
to train the Rams football team, then to Monaco for the Formula One
races. The next year included visits to Real Madrid and FC Barcelona
soccer teams, as well as the NBA finals.

It wasn't hard to get famous people to use Instagram Stories. As
Systrom had seen at the Oscars, many had gotten used to sharing
behind-the-scenes content on Snapchat. And the celebrities too were
worried about growth and relevance, just like the Catholic Church.
The Formula One owners were trying to get young people into racing,
and without Instagram, few people would know what Lewis Hamilton
looked like without a helmet. Timberlake, already a household name
with about 50 million followers, could only really grow his following
by being exposed to the audience for the @instagram account, which
topped 100 million at the time.

In fact, when stars were featured prominently on @instagram, other
celebrities would volunteer to demo the product in exchange for access-
ing that audience. Once Taylor Swift's team saw other superstars featured
on @instagram Stories, they reached out, asking for the same treatment.
Chen flew out to spend time with Swift in her apartment, filming her
with her cats, to subtly teach Instagrammers that Stories was about less
polished moments.

· · · · · · · · · · · ·

Soon after Stories launched, Instagram took a symbolic step out of its
parent company's shadow. The employees moved off campus, out of
Hacker Square and into a multistory glass building about a five-minute
shuttle bus ride from Facebook's like button sign.

When Marne Levine, head of operations, first saw the space, she
thought it wouldn't fit Instagram's artsy vision, specifically Systrom's, especially in light of #trashcangate. It was full of drab cubicles. But Systrom and Krieger saw the possibilities in renovation, and accordingly,

the whole insides were stripped and reimagined with minimalist surfaces, white paint, fresh light wood, and plants. Instead of Facebook's motivational posters, pictures by Instagram users in frames were hung along the walls. There was high-quality coffee from a Blue Bottle shop on the ground floor. Right in front of the office was a large white outline of the Instagram logo, representing the first time Instagram had marked its territory so prominently.

Instead of the Gravity Room, there was a whole row of dioramas for visitors to pose in. One allowed people to float in a sunset, the pink, purple, and orange gradient backdrop evoking the colorful new app logo behind them, and giant bulbous plastic clouds in front. Other dioramas allowed photographs with a glowing planetary orb, or in the middle of a starry sky.

Employees, when given more, expected more. Levine told staff she was open to suggestions. So people sent her photos of the least Instagrammable food in the cafeteria—most egregiously, a large vat of potato salad. Someone even joked about the starchy, mayonnaisey eyesore at the Instagram leadership meeting.

Systrom was sympathetic. "In all seriousness, this is why it's important," he told Levine. "We're asking our employees to think about simplicity and craft and the community and to internalize what's important. You want the salad bar to present as if it's a crafted experience that you're excited about."

Got it, Levine thought. *It's not about potatoes. It's about our values.*

Four years had passed since Instagram had joined Facebook. Now they were negotiating with Facebook to have a major office in New York, and eventually one in San Francisco, back where everything started.

. . . .

Systrom was feeling invincible. Two weeks after the Stories launch, he recovered from the anxiety with a vacation. He was still on his cycling kick, trying harder and harder rides on different types of bikes he purchased from King. So on the trip, he challenged himself to summit Mont Ventoux, one of the most difficult hills in the Tour de France. "I have never worked as hard as I did on this climb, but I survived!" he posted on Instagram, posing triumphantly with his bike and a bottle of Dom Pérignon. His caption told the world his time was 1:59:21, only double the overall record of one hour.

With that, he had finally earned the right to order his ultimate fancy bike: a Baum.

The Australian maker of the bespoke bike would need a couple months to craft it out of custom-butted titanium to make it as lightweight as possible and tune it for Systrom's specific riding style. It would also feature specific red and blue stripes, an homage to Martini car racing, that would take thirty hours to paint. Nate King was tickled, because most of the people buying a Baum from his shop did it for the prestige; he knew that Systrom would actually ride it.

Instagram had its own billions in revenue, its own world-changing app, its own product vision and strategy, and its own offices. Its leaders had learned how to make difficult decisions by recognizing their blind spots and removing the high bar for posting. Employees allowed themselves to feel, for a few victorious months, like they might one day be as important as Facebook. A Facebook 2.0, making decisions more thoughtfully, in a way that made users happier, borrowing some lessons and rejecting others, modeling the future of social media.

They might, if they kept going in this direction, make it to 1 billion users.

But soon, Facebook would be in crisis. And Zuckerberg wasn't about to let Instagram forget whom they were working for.

CANNIBALIZATION

"Facebook was like the big sister that wants to dress you up for the party but does not want you to be prettier than she is."

—FORMER INSTAGRAM EXECUTIVE

One day in October 2016, Kevin Systrom sent a note to his policy head, Nicky Jackson Colaço, explaining that he needed a briefing document. He was going to meet with Hillary Clinton that evening at a fundraiser for her presidential campaign.

Jackson Colaço was troubled by Systrom's ask. She was a Clinton supporter herself, but Systrom was the CEO, representing Instagram in public. She wished she'd had more warning, because this would need to be handled carefully. Was he going to meet with the Republican candidate, Donald Trump, too? The world was watching—and gauging Facebook's impartiality in the coming election.

Earlier that year, as Instagram was building Stories, the online technology news site *Gizmodo* had written about a team of Facebook contractors who curated news into a "trending topics" module on the right side of the news feed. It was the only human-led editorial component of the social network. The blog cited anonymous Facebook contractors who said they routinely served up content from publishers like the *New York Times* and the *Washington Post*, but eschewed right-wing Fox News and *Breitbart*. *Gizmodo* also reported that employees were openly asking Facebook management whether they had a responsibility to prevent a Trump presidency. The reporter implied it was scary that Facebook's employees realized their company *was* powerful enough to do this, if it wanted to.

Facebook, in its response to the firestorm from the leaks, invited 16 of the most TV-friendly conservative political commentators, including Tucker Carlson, Dana Perino, and Glenn Beck, to its headquarters to learn about how the news feed was programmed. They reassured their guests that Facebook had no editorial bent. Later, they cut humans out of the process for picking trending topics, so that what trended on Facebook was determined by algorithm only.

Even after all those efforts, the company still feared an outcome that seemed likely at the time—that once Clinton was elected, everyone would blame Facebook for tilting the scales in her favor. Facebook executives didn't want to alienate the conservative portion of their U.S. users, so their pre-election strategy was to appear as equitable as possible, letting the news feed algorithm show users whatever news they wanted to see. To be extra fair, they offered advertising strategy help to both the presidential campaigns, though only the Trump campaign accepted it. Clinton's team was already experienced with running for president.

In the midst of all this, Systrom felt like Instagram was independent enough that it wasn't necessary for him to feign impartiality in the election. He told Jackson Colaço that he was entitled to his own views as a private citizen. Later that night he posted a selfie with Clinton, emphasizing in a caption that he was *personally* impressed with her: "I hope that Instagram can be a place for you to voice your support for whatever candidate you may choose. For me, I'm very excited for Secretary Clinton to be the next president of the 🇺🇸 #imwithher."

The incident would highlight an emerging chasm between the booming app and its increasingly controversial parent company. Because despite Jackson Colaço's concerns, Systrom's post made no waves. It turned

out that the public didn't think of Instagram as part of Facebook's controversy, or as part of Facebook, period. The brands were so separate
that U.S. users saw Instagram as an escape from the big social network's
political debates and viral fluff. Most Instagram users had no idea that
Facebook owned the app. Systrom and Krieger had been careful to preserve their reputation.

The debate about what news stories Facebook surfaced wasn't really
about bias so much as it was about power. Facebook had amassed unprecedented control over public conversation, in ways that were opaque to its
1.79 billion users. The company had done everything it could to grow its
network and the amount of time people spent on its site, with unintended
consequences.

Facebook had wanted to beat Twitter, and so had encouraged more
news publishers to post on the social network. The plan worked. Their
users were discussing the top news, which in the U.S. was the election.
But now Facebook was under fire for what users read, and for the fact
that the network's ultra-personalization meant each user saw a slightly
different version of reality.

Facebook had wanted to grow its users' social networks, thinking bigger
networks were more valuable to them, and would keep them logging in more.
That had worked too. But now everyone's network included people with
loose ties to their lives, like former coworkers and friends' ex-boyfriends—
acquaintances they might never keep around if not for Facebook. People
weren't posting as many personal updates as they had in years past. Instead,
they were taking quizzes about which *Harry Potter* character they were most
like, and wishing their distant contacts an obligatory "Happy birthday!" because Facebook reminded them to. And they were having conversations it
would be easy for anyone to join—about politics.

With friends not posting as much about their personal lives, Facebook found a new kind of update to stuff into the news feed: any public
post a friend commented on, even if it was from someone outside their
network. That increased the amount of virality on Facebook, because a
person didn't have to choose to share something in order for a wider audience to see it. At the company they called this an "edge story," because

it happened at the edge of a user's friend circle. Again, the move helped spread political debates on Facebook.

Instagram, unlike Facebook, actually made human-led editorial decisions frequently. But nobody called them out as biased. If the community team wanted to highlight dogs and skateboards on @instagram instead of people with nice abdominal muscles, so be it. Instagram had created what felt like a friendly alternative to Facebook, which allowed people to consume or create content related to their interests, whether ceramics, sneakers, or nail art—interests they might not have found until Instagram offered them up via their various curatorial strategies.

It was all of these things Instagram avoided—hyperlinks, news, virality, edge stories—that cheapened Facebook's relationship with its users. Facebook was indeed biased, not against conservatives, but in favor of showing people whatever would encourage them to spend more time on the social network. The company was also in favor of avoiding scandal, appearing neutral, and giving the public what they wanted. But as Facebook became a destination for political conversations, the human curation in "Trending Topics" wasn't the actual problem. It was how human nature was manipulated by Facebook's algorithm, and how Facebook looked away, that got the company in trouble.

· · · · · · · · · · · ● ● ●

Few at Facebook expected that Donald Trump would win the 2016 presidential election. At the Menlo Park campus the day after the vote, the mood was dark, employees whispering in corners and checking their phones. Some of them stayed home, too emotional to face the reality of the erratic new leader of the United States.

The media generated several handy narratives for How It Happened. A top theory was that the news feed algorithm, designed by engineers to give people what they wanted, had been rewarding articles and videos that subtly nudged voters to believe in outlandish conspiracy theories and fake news that often cast Clinton herself in the worst possible light.

Stories that claimed the pope had endorsed Trump, or that Clinton had sold weapons to the Islamic State, were juiced by Facebook's algorithms and promoted to millions of Facebook users. In the three months prior to the election, the top stories with false information reached more people on Facebook than the top stories from legitimate news outlets. Some of them came from makeshift websites designed to look real, with names like *The Political Insider* and *Denver Guardian*. On Facebook, the subterfuge worked. All links were presented in identical fonts on the news feed, awarding a scrappy conspiracy theorist the same credibility as a fact-checked report from ABC News. One such site even had the web address ABCnews.com.co, even though it was not affiliated with the network.

The most shareable content on Facebook was what made people emotional, especially if it triggered fear, shock, or joy. News organizations had been designing more clickable headlines ever since the social network became key to their distribution. But those news organizations were getting beaten by these new players, who had come up with an easier, more lucrative way to go viral—by making up stories that played on Americans' hopes and fears, and therefore winning via the Facebook algorithm.

The fake news on the websites would be interspersed with stories that weren't entirely fake, but were hyper-slanted with context or loaded language designed to reaffirm readers' paranoia or political loyalty. People shared these stories to prove to their friends and families that they'd been right about everything all along. Meanwhile, the slanted sites built advertising businesses off the traffic they were getting from Facebook.

Some Facebook executives, like Adam Mosseri, the head of the news feed, had been ringing alarm bells about misinformation internally, wanting to make it against the social network's content rules. But Joel Kaplan, the vice president of public policy, who was a political conservative, thought that kind of move could be dangerous for Facebook's already tenuous relationship with Republicans. A lot of the incendiary stories benefited Trump, and in removing them the company would have fueled those fears regarding Facebook's bias.

The day after the election, with employees still shell-shocked, Elliot Schrage, the head of policy and communications, convened with Zuckerberg and Sheryl Sandberg, and decided that Facebook's role in the vote was getting unfairly overplayed by the media. They needed to address the criticism. Facebook was simply creating a digital hangout space—a neutral zone like a town square, where anyone could say what they wanted to say, and be corrected by their friends if they were wrong. The trio came up with a defensive messaging posture, pushing the idea that freethinking American citizens made their own decisions. Zuckerberg, at a conference just two days after the election, said, "I think the idea that fake news on Facebook—of which it's a small amount of content—influenced the election in any way is a pretty crazy idea."

The comment drew immediate ire, because the public now realized the news feed algorithm had the power to shape what citizens understood about their candidates. If Facebook users were all in a digital town square, they were each listening to the public speaker Facebook thought they would find most interesting or urgent, while experiencing whatever companions and entertainment Facebook thought would please them. Then, without any knowledge of what someone else's town square looked like, users were trying to make a decision collectively about who the mayor should be.

But Zuckerberg gave the same dismissive opinion at a question-and-answer session with employees the next day. He also told his workers that there was another, more positive way of looking at it. If people were blaming Facebook for the election's outcome, it showed how important the social network was to their everyday lives.

Not long after Zuckerberg's talk, a data scientist posted a study internally on the difference between Trump's campaign and Clinton's. That was when employees realized there was another, maybe even bigger way their company had helped ensure the election outcome. In their attempt to be impartial, Facebook had given much more advertising strategy help to Trump.

In the internal paper, the employee explained that Trump had outspent Clinton between June and November, paying Facebook $44 million compared to her $28 million. And, with Facebook's guidance, his campaign had operated like a tech company, rapidly testing ads using Facebook's software until they found the perfect messaging for various audiences.

Trump's campaign had a total of 5.9 million different versions of his ads, compared to Clinton's 66,000, in a way that "better leveraged

Facebook's ability to optimize for outcomes," the employee said. Most of Trump's ads asked people to perform an action, like donating or signing up for a list, making it easier for a computer to measure success or failure. Those ads also helped him collect email addresses. Emails were crucial, because Facebook had a tool called Lookalike Audience. When Trump or any advertiser presented a set of emails, Facebook's software could find more people who thought similarly to the members of the set, based on their behavior and interests.

Clinton's ads, on the other hand, weren't about getting email addresses. They tended to promote her brand and philosophy. Her return on investment would be harder for Facebook's system to measure and improve through software. Her campaign also barely used the Lookalike tool.

The analysis, which wouldn't leak to Bloomberg News until 2018, proved that Facebook's advertising tools, when used in the right way, were extremely effective. Trump's win, in part because his team had taken full advantage of Facebook's power to personalize and target information to a receptive audience, was an ideal outcome for any top advertising client. But Trump hadn't been selling cookware or flights to Iceland—he'd been selling the presidency. So the customer success didn't make the mostly liberal workforce feel any better. Zuckerberg had always told them they were going to change the world, making it more open and connected. But the bigger Facebook got, the more it had the power to shape global politics.

A few days later, at a gathering of world leaders in Lima, Peru, President Barack Obama tried to say as much to Zuckerberg. He warned the CEO that he needed to get a handle on how Facebook spread falsehoods, or else the misinformation campaigns in the 2020 presidential election would be even worse. Obama knew from U.S. intelligence, but didn't say to Zuckerberg at the time, that some of the incendiary news wasn't coming from shady media entrepreneurs. One of the country's biggest adversaries was running a pro-Trump Facebook campaign too.

Zuckerberg reassured the outgoing president that the problem was not widespread.

It was an uncomfortable time for Instagram to be thriving. As Facebook executives were still agonizing over how to avoid blame for the election result, Systrom presented a plan to increase head count for the team working on Instagram Stories. Stories was still a simple product, but already very popular. Systrom saw opportunities to add more features to it, like face masks and stickers similar to those offered by Snapchat.

Michael Schroepfer, his manager and Facebook's chief technology officer, denied the request. "You should just pivot the team you have to work on Stories," Schroepfer said. "We want to see that you're making hard trade-offs before allocating any new people." Facebook was working on its own versions of posts that disappeared after 24 hours, not just for Facebook, but for WhatsApp and Messenger—all of which would be slightly different than Instagram's version.

Other managers, from other parts of the company, found Schroepfer's resistance unusual. Why not reward Instagram for its success? Why was Facebook making other versions of Instagram's Stories product, instead of just throwing its support behind Instagram's? Other teams, in virtual reality, in video, and in artificial intelligence, were having no trouble getting head count. The team of people working on Facebook Stories was already quadruple the size of Instagram's.

But Krieger and Systrom chalked it up to history. Instagram had always made things work with a smaller team than competitors. Perhaps Facebook reasoned that they worked better that way. In the weeks that followed, Systrom fought for more people, and eventually got some help. But the experience was a harbinger of the problems ahead.

· · · · · · · · · · · ● · ·

Zuckerberg was absorbed in some issues that had nothing to do with the U.S. presidential election—and, in fact, concerned him more. Even though Facebook was still growing, the way people were using it wasn't trending well for its future. And Instagram Stories wasn't going to solve that.

The first problem was about how Facebook fit into its users' days. While people spent an average of about 45 minutes per day on Facebook, known internally as the "big blue app," they were doing so in short sessions—an average of less than 90 seconds per sitting, according to an internal data analysis. They were not lounging with Facebook on their couches so much as they were checking it at bus stops, in line for coffee, and on toilet seats. That was a problem if Facebook wanted a bigger chunk of the most valuable advertising market: television.

Facebook had been prioritizing video in the news feed algorithm, even promoting live videos, but the kinds of things that dominated were quick viral clips that caught users' attention as they scrolled through their news feeds. They'd stop and check out a video of an adorable puppy or a funny stunt, but since they weren't actively picking the content, they often wouldn't watch long enough to see an advertising break. The videos that got the most traction were often low-quality, produced or repurposed by content farms, with networks of Facebook pages that would promote whatever they posted to make it go viral. There were few Facebook "creators" like the ones who had become famous building audiences on YouTube and Instagram.

So Facebook's temporary solution to get users to watch long-form videos, and therefore video ads, was to create a new premium part of the social network just for that kind of content. Facebook would pay TV studios to make the higher-quality shows its users weren't. The video site, which would eventually be called Facebook Watch, was a solid plan to take on television and YouTube, and solve Zuckerberg's first problem.

But there was a second problem. People on Facebook were not posting updates like they used to. They were sharing links and making events, but they weren't posting their original feelings and thoughts as often. Earlier that year, Facebook had tried to make it more fun to post, giving users options for colorful backgrounds and fonts so their writing would become more eye-catching. The social network was even prompting people on the top of their news feeds with photos from their pasts, so perhaps they would re-share memories. Facebook was also alerting them about obscure holidays and events, like National Siblings Day, in hopes that people would post about them.

Adding a disappearing-posts product across Facebook properties was one way to solve the issue, by lowering the anxiety associated with posting permanently, just as Instagram had done. But Zuckerberg, ever paranoid, wondered if that was enough.

He looked at Instagram's growth, which was actually accelerating, even as the rate that Facebook, Twitter, and Snapchat added users was slowing. The discovery didn't bode well for his prized acquisition.

Zuckerberg reasoned that Facebook's users only had a certain amount of minutes in their day, and it was his job to get them to spend as many of those spare moments on Facebook as he could. Maybe the problem wasn't just that they were lured away to Snapchat or YouTube. Maybe the problem was that all of his users had an alternative social network to visit—one that Facebook was promoting on its own site, and had been for years.

· · · · · · · · · · · · ● ● ●

As the other Facebook clones of Snapchat Stories started to roll out, none of them made as big a splash as Instagram's. The Messenger chat app started testing the feature in September, calling it "Messenger Day." Then Facebook tested Stories on the main app in January, calling them Stories too. Even WhatsApp added similar functionality in February, calling it Status, after Zuckerberg pushed for the move in a heated battle with the app's founders. Now the public had four different Facebook-owned but separately branded places to post disappearing video to their friends, just like they could on Snapchat.

Zuckerberg was willing to try multiple things at once to quash competitors. But having all the options was confusing, not exciting, for the public. People didn't understand why they needed the new features, or which of their friends had access to them and which didn't. And there was no exciting celebrity content to train them on what to do, like employees had made for @instagram.

As *The Verge* wrote at the time, "borrowing Snapchat's ideas is working out okay for Instagram, but for some reason Facebook's direct attempts always feel a little off—and desperate."

Zuckerberg didn't see the matter in terms of "feel." He saw it in terms of Instagram stealing Facebook's opportunity.

He told Systrom, over the course of multiple meetings, that he thought Instagram was successful with Stories not because of its design, but because they'd happened to go first. If Facebook had gone first, perhaps Facebook would have become the destination for anyone who wanted that kind of ephemeral experience. And that might have actually yielded a better outcome for the overall company. Facebook, after all, had more users and a more robust advertising operation.

Systrom hadn't expected this kind of feedback. Going first might have helped Instagram with its cool factor, but if moving first was all that mattered, there would be no reason to copy Snapchat. Facebook might have purchased Instagram as a defensive strategy, but if his team was taking shots and scoring, why was that a bad thing? He was winning, but it felt like losing. It seemed as if, in the order of priorities, a win for Facebook the Social Network was more important than a win for Facebook the Company.

But Systrom didn't argue. He'd seen Zuckerberg fight with other, more headstrong leaders at Facebook, especially from acquired companies WhatsApp and Oculus, the virtual reality arm, and knew how it could end. For example, after Zuckerberg bought Oculus in 2014, he wanted to change the name of their virtual reality headset, the Oculus Rift, to the Facebook Rift. Brendan Iribe, a cofounder of Oculus and then CEO of that division, argued that it was a bad idea because Facebook had lost trust with game developers. Over a series of uncomfortable meetings, they settled on "Oculus Rift from Facebook." In December 2016, after a number of similar disagreements, Zuckerberg pushed Iribe out of his CEO position.

When someone is having an emotional reaction, you don't poke at it, Systrom thought. And anyway, Systrom, with Taylor Swift's help, was already working on what was supposed to be Instagram's next bold idea.

• • • • • • • • • • • • •

Systrom wanted to capitalize on the idea that Instagram was an escape from the rest of the internet, a place where things were more beautiful and people were optimistic about their lives. The biggest threat to that brand, which Ariana Grande and Miley Cyrus had raised in years past, was the fact that on an anonymous network, it's easier for people to say hateful things about one another. Finally, Systrom decided, it was time to take on bullying.

But in Instagram style, the plan started out as a reaction to a celebrity, this time Taylor Swift, in a crisis on the site. (Instagram might have resolved to prioritize its regular users in product builds, with changes like the algorithmic feed, but it was still listening intently to celebrities about their needs, reasoning that doing so was good for the brand, as the celebs' problems also affected their millions of followers.) The pop star, who knew Systrom through close friends, the investor Joshua Kushner and his supermodel girlfriend Karlie Kloss, started having a major problem that summer before the election. The comments on her photos were being bombarded with snake emoji, and the hashtag #taylorswiftisasnake.

She was in two public disputes with other celebrities. After she split with her boyfriend, producer and DJ Calvin Harris, Swift revealed that she'd helped write his hit song with Rihanna, "This Is What You Came For." The revelation dominated the coverage of the song. He didn't appreciate Swift making him look bad post-breakup, saying it had been her decision to use a pseudonym in the credits. Fans of Rihanna and Harris started calling her a snake—a sneaky person.

In a separate incident around the same time, she criticized Kanye West for the lyrics about her in his song "Famous," debuted that February 2016: "I feel like me and Taylor might still have sex / I made that bitch famous." Kim Kardashian West retaliated in a July episode of *Keeping Up with the Kardashians*, where she shared a video of a conversation between Swift and her husband on Snapchat. In the video she gave approval for the "might have sex" part of the lyrics (though the "made that bitch famous" part remained up for debate).

On a day that was apparently National Snake Day, Kardashian West tweeted: "They have holidays for everybody, I mean everything these days," followed by 37 snake emoji, in a veiled reference to Swift. The reptile takeover of Swift's Instagram page accelerated.

Swift's team had a close relationship with Instagram's. Once, Charles

Porch, the head of partnerships, had given them a heads-up about a hack of her account before they realized it. So they asked if there was anything Instagram could do about the snakes. Systrom wanted to automatically delete all the reptilian vandalism en masse. But people would notice. Jackson Colaço made the point that they couldn't make a tool just for a famous person, without making it available to everyone else.

Swift wasn't the only one feeling like her Instagram comments had been taken over by anonymous haters. Around the same time that summer, Systrom and Krieger made their first trip to VidCon, a conference where famous people on the internet connect with partners and studios. Hordes of teens and tweens make their parents bring them to the event too, trying to catch a glimpse of their favorite digital stars. It takes place in Anaheim, California, next door to Disneyland. Systrom and Krieger hosted an after-party at the Disneyland Dream Suite, an exclusive apartment within the theme park where Walt Disney himself used to live.

Many of the stars, known by the term "creators," explained that their Instagram pages were regularly vandalized by internet trolls. On Instagram, everything they did was carefully curated. Their posts weren't just for alerting followers to new YouTube videos, they were supposed to demonstrate to brands how positive it would be to work on a sponsorship campaign together. And these days, brands were looking at comments to understand their return on investment.

After Systrom had been convinced of the product opportunity, the team developed a tool to hide comments by filtering out a specific emoji or keyword, which anyone could use, not just Swift. It served as a major relief, especially to people with thousands or millions of followers, for whom it was untenable to delete comments one by one. When Instagram finally talked about the tool's origin story months later, they framed Swift as a "beta tester" helping the company out. They protected the fact that she'd been bothered by the onslaught.

Systrom decided Instagram should lean into its feel-good image, giving people even more tools to block out what they didn't want to see. By December 2016, Instagram was letting users turn off comments for posts entirely if they wanted. Systrom's willingness was in stark contrast to the attempts by Facebook and Twitter to err on the side of leaving content up, in an attempt to promote environments they said were neutral and open, but that in practice were rarely policed.

The same ideas, of letting users turn off comments or block them according to keyword, had been suggested several times at Facebook over the years. But it had never stuck. If there were fewer comments, there were fewer push notifications, and fewer reasons for users to come back to the site. Even on Instagram's team, the former Facebook employees promised Systrom that they would find a way to build out the tool so it was difficult to find, and applicable only to one post at a time. That way, it wouldn't be used as often.

Thanks but no thanks, Systrom said. He explained that he wasn't worried about losing engagement, that the team was thinking too short-term. Over the long term, if the tool was easy to find and well publicized, people would have more affinity for Instagram, and the product would better weather storms of bad publicity, like the kind Facebook was starting to receive.

Systrom wanted to think even bigger than comments. He started talking to Jackson Colaço about a "kindness" initiative. How could Instagram, with its more aggressive editorial approach, giving more power to its users, become the internet's utopia?

· · · · · · · · · · · · ·

Zuckerberg, meanwhile, had high hopes for how people would view Facebook. Yes, Facebook was powerful, but the election had proved it was also much maligned. If only the public could see Facebook as he did, as a tool to create empathy in the world, not division. He was on a mission to reframe his giant network as a humanitarian project.

Critics were still saying Trump's election and Britain's vote to leave the European Union were the result of a Facebook-fueled polarization of society. One of Zuckerberg's least favorite criticisms of Facebook was that it created ideological echo chambers, in which people only engaged with the ideas they wanted to hear.

Facebook had already funded research, in 2015, to show echo chambers were mathematically not their fault. With the social network, everyone had the *potential* to engage with whatever kinds of ideas they wanted to, and tended to have at least some Facebook connections with people who held different political opinions. But if people *chose* not to interact with those they disagreed with, was that really Facebook's doing? Their algorithm was just showing people what they demonstrated, through their own behavior, they wanted to see, enhancing their existing preferences.

Zuckerberg felt he needed to explain to the public that Facebook could be a force for good. And so he wrote a 6,000-word manifesto, which he posted to his Facebook account in February 2017. "In times like these, the most important thing we at Facebook can do is develop the social infrastructure to give people the power to build a global community that works for all of us," he wrote. He used Instagram's favorite term, "community," 130 times, but didn't say much of actual substance about what Facebook would build. Either way, Zuckerberg seemed to be saying that whatever problems Facebook had been blamed for creating, their solutions would be built by Facebook.

Zuckerberg backed up his promise with a commitment to better understand his users. For more than a decade, he'd been the CEO of Facebook, thinking about keeping the product alive and thriving, meeting with employees, other CEOs, and world leaders. He didn't often get to meet regular people.

So he decided to make a new year's resolution. Doing so was an annual tradition for him: in 2011, the year before he acquired Instagram, he announced he would eat only meat he killed himself. In 2016, he determined that he would build his own home artificial intelligence assistant. In 2017, he wanted to visit all 50 states in America in an attempt to understand a broader swath of his user base.

Zuckerberg, often with his wife, Priscilla Chan, who co-ran the family philanthropic investments, completed his challenge by visiting farms, factories, and diners to meet regular people. But his efforts to mingle with the everyman were often thwarted by how orchestrated the visits were. Much more planned than Systrom's meetings with Instagram users, at times Zuckerberg's stops resembled a presidential candidate on tour.

A staffer would tell the host in the selected state that someone important—a Silicon Valley philanthropist—was coming to visit. Then Zuckerberg's security team, made up of former U.S. Secret Service agents, would sweep the location. With an entourage that included a professional photographer, so he could later post the pictures on his Facebook page, Zuckerberg would arrive. A communications team would help write and edit his speeches and Facebook posts, which were always a combination of heartfelt anecdotes, insights on humanity, and dad jokes.

The tour wasn't having a positive effect on the leader's reputation or the social media platform. Since at least 2015, Facebook's communications team had regularly polled the public, surveying users on Facebook about whether the company was innovative and good for the world. Considering Zuckerberg's reputation and that of the company were inextricably linked, they also asked these questions about Zuckerberg himself. Because Zuckerberg's tour wasn't helping the numbers, Facebook's communications team gathered for an off-site meeting that spring of 2017, where Caryn Marooney, the PR head, presented research showing Facebook's brand had a more unfavorable rating than that of Uber—a ride-sharing startup beset with scandal during that period.

Emily Eckert, Systrom's business lead, shot a look at Kristina Schake, the head of comms at Instagram. "Should I ask if they polled about Instagram's brand?" she whispered, thinking it would be funny to make the room instantly uncomfortable about the discrepancy.

Kristina shook her head, smiling. "Don't you dare!"

• • • • • • • • • • • • •

While Zuckerberg was crafting his outreach, the Facebook leadership team was conducting a routine product review with Instagram's senior executives. The Instagram founders' usual goals for these meetings were to come in, let the higher-ups at Facebook Inc. know their plans, and get the minimal approval and feedback they needed to proceed. Zuckerberg usually had a couple sentences to give, often a comment about something Instagram could try in order to achieve better growth. But obtaining his approval was like checking a box, after which Systrom and Krieger could get back to running their company as they liked.

The founders still thought of Instagram as its own company, despite all the integrations and resources from Facebook. This low-stress review process was part of the reason. It's what Systrom was referring to in his media interviews when he'd say Zuckerberg was more of a board member to Instagram than a boss.

This time, Krieger and Systrom felt they'd done exactly what Facebook wanted. By the time of the meeting, they had 600 million users: they were on their way to having a coveted 1 billion, if things kept trending as expected. They were contributing billions in revenue, with the help of Facebook's ad technology.

But Instagram got a much more intense review than they'd anticipated. Zuckerberg explained that he had some major concerns, using a word that evoked violent imagery and alarm: "cannibalization." The CEO wanted to know, if Instagram were to keep growing, would it start to eat at Facebook's success? Wouldn't it be valuable to know if Instagram was going to eventually siphon off attention that should be allocated to Facebook?

The questions lent insight into how Zuckerberg viewed his users' choices. The discussion wasn't about whether people preferred to be on Instagram versus Facebook. Their behavior was malleable. Facebook knew exactly how much traffic they were sending to Instagram through their app. They knew exactly which ways the parent company was supporting the growth of its acquired company, through links and promotion on Facebook. And if they found that Instagram's growth could be a problem for the main application, they could find a way to *fix* that.

First, they needed to analyze the problem. Alex Schultz on the Facebook growth team was asked to look into cannibalization, with help from about 15 data scientists at both Facebook and Instagram.

· · · · · · · · · · · ·

By April 2017, Zuckerberg's pronouncements about global community had started to feel more like preemptive strikes in a public relations battle.

That month, Facebook put out a cryptic research paper explaining that it had found instances of "information operations" conducted by "malicious actors" on its social network. Basically, some entities (they didn't say who) were creating fake identities on Facebook, befriending real people and spreading misinformation, trying to skew public opinion. As Obama had warned, the fake-news problem wasn't just about a few shady entrepreneurs—it was about foreign actors that had weaponized the social network's algorithm.

This revelation played into political suspicion around whether Russia had assisted in Trump's win, and whether all of those fake stories—about the pope endorsing Trump, or about Clinton working with the Islamic State—had gotten to be so popular on social media as part of an elaborate propaganda campaign. Was Facebook a part of Russia's plan? Was Russia helping Trump?

Facebook wouldn't say, arguing that it would be irresponsible to do so. Russia was the company's hypothesis, but it was a big deal to outright accuse the leadership of an entire country with millions of Facebook users in it, and what if they were wrong somehow? For months, Facebook was confident enough to assert clearly that Russian money was *not* involved. "We have seen no evidence that Russian actors bought ads on Facebook in connection with the election," the company told CNN in late July.

Naturally, this was frustrating for Democrats, who were trying to build an understanding of Russia's involvement in Trump's win. They pushed Facebook first behind the scenes and then publicly, until in September, Facebook contradicted itself, revealing for the first time that not only had Russia been behind the propaganda campaign they'd mentioned in April, but they'd purchased ads to promote their posts. Facebook had accepted at least $100,000 in advertising revenue from fake users, acting on behalf of a foreign power, because of an easy-to-use advertising system that allowed anyone with a credit card to make a purchase.

Thus began a period of public reckoning—of congressional hearings and promises and apologies, as well as more revelations and media bombshells. Twitter and YouTube revealed similar bouts of election-related propaganda from Russia. Instagram, meanwhile, cashed in on a fortuitous side effect of its acquisition by Facebook. It enjoyed the massive support and scale of the social network but was an afterthought in the rancor that often surrounded it.

When Facebook's chief counsel, Colin Stretch, testified in front of the Senate Intelligence Committee on November 1, 2017, alongside lawyers from Google and Twitter, he revealed the most troubling statistics yet about Russia's influence on the election. More than 80,000 posts from Russian accounts had been posted to Facebook, some boosted by advertising, stirring up controversy in the U.S. about immigration, gun control, gay rights, and race relations. Russia's goal had been to infiltrate interest groups in the United States, and then make them angry. In the process, Stretch said, the posts had gone viral, reaching 126 million Americans.

Later in the hearing, a senator asked about Instagram specifically. "The data on Instagram isn't complete," Stretch said. But he estimated that Russian posts on Instagram had reached an estimated 16 million people; Facebook later revised that number to 20 million. So the true reach of Russia's campaign on Facebook-owned properties was more like 150 million. Instagram got to be an afterthought in the conversation.

Eventually, Zuckerberg, Sandberg, and other Facebook executives, including Chris Cox, the head of the Facebook app; Schroepfer, the CTO; and Monika Bickert, the head of policy, would represent the company in public testimony in front of various governments around the world, as more scandals unfolded. So would Twitter CEO Jack Dorsey and Google CEO Sundar Pichai.

But Systrom was never asked to testify. And journalists continued to write about his plan for an internet utopia, presenting him as the more thoughtful social media executive.

With Instagram's biggest problems in Facebook's hands, he had the luxury of avoiding blame. Instagram's advertising, including all the ads from Russia, was run through Facebook's self-service system. Facebook's operations team was in charge of scanning all the rule-breaking content, including on Instagram. Jackson Colaço and a couple others stepped in to help Facebook with its investigation whenever Facebook asked. But mostly, for the Instagram employees, ignorance was bliss.

•••••••••••• • ••

While Facebook was concerned with the election fallout, Systrom was preoccupied with analytics. Data was religion at Facebook, but never provided a perfect picture in terms of user behavior. It could tell you what people were doing, but not necessarily why.

The Instagram Stories product, for example, was disproportionately popular in Spain after it launched. Employees in analytics found out the reason why only after asking their European colleagues on the community team. It turned out that younger people were using the tool to play an alluring game, which started when someone would direct-message their friend a number. That friend would then use that number to say a secret thought about the messenger publicly ("#12 es muy lindo!") in their disappearing stories.

In Indonesia, Instagram's data analysis caught what it thought was a massive spam ring: people were putting up photos and then taking them down quickly. But once Instagram investigated further, they found the activity wasn't nefarious; it was just a sign that people in the country were starting to use Instagram for online shopping. They were posting photos of products for sale, then deleting the pictures once they were sold.

Another spam filter, which automatically suspended users who were posting a certain number of comments per minute, ended up blocking teens chatting with their friends, who had a higher frequency of activity on the app than Instagram had planned for when designing the automatic suspension to curtail spam.

So Systrom understood the limitations of just numbers, which was one reason he'd invested so heavily in direct outreach and research. But now that Facebook was studying whether Instagram was statistically likely to cannibalize Facebook's success, he wanted to get better at forecasting.

Systrom read a stack of books and talked to Mike Develin, head of analytics at Instagram, to try to understand the factors involved in making reasonable predictions for products. One night around dinnertime, he messaged Develin, saying he'd come up with a time-spent estimate for Instagram for the second half of 2017. He expected each user would spend about 28 minutes a day on the app.

Systrom's methodology for reaching the number wasn't crazy, Develin thought: *If I were teaching an undergraduate course on forecasting, this would be a very reasonable homework assignment and that answer would probably get a good grade.* His team came up with a much more scientific forecast, which wasn't far off from Systrom's number.

Systrom wasn't trying to do Develin's job. He'd just taken up analytics, the way he'd started to learn cycling. He wanted to understand the process better, so he would be prepared to parse whatever happened next.

· · · · · · · · · · · **·** · ·

At the end of Facebook's most challenging year, the social network and its smaller photo-sharing subsidiary were on a collision course. What did Instagram owe to its all-powerful parent company, which had helped Instagram reach the masses, but was now concerned about staying on top?

Systrom thought that Instagram owed Facebook continued success as a company within the company. That way, even if Facebook's troubles continued, there would be an alternative fast-growing network for anyone to visit to catch up on their friends and family. Instagram could one day be critical to Facebook's longevity—maybe it could even be the dominant platform. That year, they had helped disarm a major competitor, Snapchat, which had gone public in March as Snap Inc. Snap shares had declined, losing almost half their value, in part on concerns that the company wouldn't be able to compete with Instagram after the Stories launch.

But in Zuckerberg's opinion, Facebook Inc. was threatened if Facebook itself wasn't thriving. Facebook was in a tough spot of its own making, dealing with more public scrutiny and skepticism than it had ever received. He had given so much freedom and support to Instagram. Now it was time for Instagram to start giving back.

When Schultz completed his research on whether Instagram would cannibalize Facebook, the leaders read the data very differently.

Zuckerberg thought the research showed that it was likely Instagram would threaten Facebook's continued dominance—and that the cannibalization would start in the next six months. Looking at the chart years into the future, if Instagram kept growing and kept stealing users' time away from Facebook, Facebook's growth could go to zero or, even worse, it could lose users. Because Facebook's average revenue per user was so much higher, any minutes spent on Instagram instead of Facebook would be bad for the company's profitability, he argued.

Systrom disagreed. "This is not Instagram taking away from the Facebook pie to add to the Instagram pie," he said in their Monday morning leadership meetings. "The total pie is getting bigger." It wasn't just Instagram versus Facebook. It was all of these Facebook properties versus every other choice in the world, like watching television or using Snapchat or sleeping.

Others in the room for those discussions were puzzled. *Has Mark forgotten he owns Instagram?* Zuckerberg had always preached the idea that Facebook should reinvent itself before a competitor got the chance, and that the company should make the decisions about how to do so based on data. "If we don't create the thing that kills Facebook, someone else will," the booklet passed out at employee orientation says.

But Facebook was Zuckerberg's invention. And in this case, the CEO was reading the data with an emotional bias.

His first decree, at the end of 2017, was a small one, barely noticeable to users.

He asked Systrom to build a prominent link within the Instagram app to send his users to Facebook. And alongside the Facebook news feed, in the navigation to all of the social network's other properties, like groups and events, Zuckerberg removed the link to Instagram.

THE OTHER FAKE NEWS

> *"It used to be that the internet reflected humanity, but now humanity is reflecting the internet."*
>
> —ASHTON KUTCHER, ACTOR

After Instagram changed its feed using algorithmic ordering in June 2016, gradually anyone using the app for promotional purposes realized that they would need to completely revise their strategy. The new feed order—which prioritized users' closest relationships instead of the newest posts—meant influencers and businesses could no longer grow their followings by simply posting often.

It was as if every fledgling Instagram-based business had the same job under a new, mysterious boss, with no insight as to why their performance was suffering. Some failed by applying the same strategies they had used in 2015. Others, via various memes and pleas to their followers, accused Instagram of robbing them of growth that was rightfully theirs. They were desperate because while the accounts were digital, they were backing real-life jobs and businesses. One of the first prominent Instagram-born companies, Poler, an outdoor gear designer known for making sleeping bags people could walk around in (#campvibes #vanlife

#blessed), eventually declared bankruptcy after failing to meet its pro-
jected growth targets.

Like its other digital rivals, Instagram didn't have a customer service number that businesses could call to discuss their uncertainties and
complaints. The people running the accounts just attempted to get to
know and understand the new rules, scouring data about their followers'
activity for clues. They pieced together that the new algorithm weighted
a post higher if other people started talking about it right away, with a
multiple-word comment, which was better than just a heart or smiley-face
emoji.

In the confusion, Instagram's most popular users had a clear advantage over others, since many celebrities and major influencers, especially
in the United States, already had relationships with Charles Porch's partnerships team. That team paid special attention to the Kardashian-Jenner
family, as they were now keepers of five of the top 25 Instagram accounts.
Almost a year after the algorithm change, in May 2017, Kim Kardashian
West would become the world's fifth person to pass 100 million Instagram
followers, after Ariana Grande, Selena Gomez, Beyoncé Knowles-Carter,
and the soccer star Cristiano Ronaldo. The Kardashian family wasn't influential enough to make Instagram undo the algorithm change, but they
were successful with a different request.

Every day, the family followed a schedule, posting about whichever
product launch or life event was their designated big news. When they
commented on each other's photos, appearing publicly supportive, it had
the added benefit of sending a strong signal to the algorithm: *this post is
important and should be ranked higher.* It was a problem because, as they
told Porch's team, the public never saw their extra efforts. The comments
section for the Instagram-famous was such a constant stream of activity
that important stuff got buried. And if you were Kylie Jenner, getting
hundreds of comments in minutes after posting about lipstick, there was
no way to see and react to a supportive message from half-sister Kim in a
way fans would expect.

Porch's team liaised with Instagram's engineers and came up with a
solution: algorithmic ordering for comments too. Starting in the spring

of 2017, comments on anyone's photos from people who were important to them—maybe they were closer friends, or had a blue checkmark by their account saying they were "verified" as a public figure—appeared positioned higher and more prominently in the display.

. . . .

So once again, as with Taylor Swift's bullying complaint, Instagram changed the product for everyone based on the feedback of a few, standing firm on their overall assessment that the algorithm helped regular users see what they most wanted to see. They'd patched one problem, but now that hundreds of millions of users and businesses depended on the app, this change had a ripple effect in ways that Instagram hadn't predicted.

Everyone with a blue checkmark, after realizing that their comments would be prominently displayed, had an incentive to comment more. The comment ranking helped brands, influencers, and Hollywood types fight their deprioritization by the main algorithm. Instagram commenting became marketing, or, in the vernacular of Silicon Valley engineers, "growth hacking."

The "hacking" didn't end there. The most strategic Insta-famous weren't just commenting on their friends' posts, but on accounts that might make them seem more well-connected and relevant than they actually were. One influencer with a verified account, Sia Cooper, @diary ofafitmommyofficial, told *Vogue* she gained 80,000 followers in just a few weeks by lovingly commenting on Kardashian-Jenner posts, though she didn't actually know the family: "I choose to comment on the highest followed accounts because this means my comment is more likely to be seen by many more users," positioned at the top of the stack. Today, she has more than 1 million followers, and has inspired others to use the same strategy.

As soon as the algorithm started prioritizing the verified comments, the media did too. Seemingly spontaneous, candid celebrity banter—stars defending themselves from critics, promoting products, or simply interacting—became fodder for entertainment news. A couple months after the comments algorithm changed, the singer Rihanna commented on a makeup company's post, criticizing their lack of foundation shades for black women. She made news for calling out racism, and her own makeup brand, Fenty Beauty, benefited. The celebrity "clapback" in Instagram comments became such a common tactic that entertainment websites started ranking and listing the most memorable ones. Emma Diamond and Julie Kramer, two Syracuse University sorority sisters who'd befriended each other in a group chat about the Kardashians, started an account after graduation called @commentsbycelebs, to highlight the A-lister commentary in screenshots before anyone else. Today they have 1.4 million followers, and enough revenue through sponsored content from clients like Budweiser that they don't need other jobs. As the algorithms shifted, so did the kinds of businesses that could build off them.

And not just the legitimate businesses. Accounts with blue verified checkmarks, still difficult to obtain in countries where there were fewer Instagram employees coordinating, became more vulnerable to takeovers by hackers. The hackers would figure out a way to break through the login and then sell the accounts on the black market, where they were becoming ever-more valuable, in part because the checkmarks made them more visible in Instagram comments.

· · · · · · · · · · · ·

The 2016 election was a turning point in public thinking about social media, and the ways people and governments can use its powers for ill. One question percolated above all others: How much should technology companies work against human nature? When their users *chose* to read hyper-partisan news, *chose* to share conspiracy theories about vaccines causing autism, *chose* to share racist tirades or the manifestos of mass shooters, what was the company's responsibility, if any, to curtail them? Facebook, YouTube, and Twitter were questioned by regulators about their policies governing users, and what kind of content they should restrict or more closely police. Representatives of each explained that they wanted to err on the side of promoting free expression and limiting takedowns, embracing the solution that happened to be cheapest, necessitating the least human oversight.

Instagram's comment algorithm change was a tiny tweak, with mostly benign effects. Self-promotion is hardly a threat to democracy or medical truth, and may have made Instagram more fun, especially through accounts like @commentsbycelebs. But the change, and its corresponding effect on user behavior, illustrated something fundamental, ignored in the arguments over content policy. Social media isn't just a reflection of human nature. It's a force that *defines* human nature, through incentives baked into the way products are designed.

Instagram measures follows, likes, and comments. Since users know they will be judged in each category with every post, they tailor their behavior to meet the standards their peers are hitting, the way a gymnast knows they will be evaluated on the difficulty and execution of their routine. The bigger Instagram grew, the more its users strived for followers, likes, and comments, because the rewards of achieving them—through personal validation, social standing, and even financial reward—were tremendous.

An Instagram user's path to success was obvious, based on benchmarking against others. All you had to do was create the right kind of content: visually stimulating, with a reflective but optimistic caption, inspiring some level of admiration. En masse, those activities spilled over into real life and real business decisions. The version of Instagram that the founders had set out to create, one that would foster art and creativity and provide visual windows into the lives of others, was slowly being warped by the metrics Instagram prioritized, turning the app into a game that one could win.

The effect had already played out in other parts of the internet, where user-generated content reigns. On YouTube, the site's algorithm gradually started to reward creators according to watch time, thinking that a longer time spent on a video meant it was engaging enough to be displayed higher in searches and recommendations. In response, those seeking fame on the site stopped making short skits and started making 15-minute makeup tutorial videos and hour-long debates about video game characters, so they could be displayed in rankings more prominently and slot in more ads. YouTube also measured average percentage of a video viewed, as well as average watch duration, as signals for ranking. So YouTube creators tailored their behavior to those metrics too, getting angrier and edgier in their videos to retain viewers' attention. Some of them stoked conspiracy theories, saying anything sensational enough to keep people tuned in. Anyone who erroneously believed in chemtrails or the flat earth found new support and community on the site.

Companies try to intuit what a good measure of happiness for their users might be and, by building their sites to prioritize those metrics, manipulate their users over time. On Facebook, once the company started rewarding its employees if they increased the amount of time users spent on the app, users started seeing more video and news content in their feeds. As was apparent in the election, rewarding content that sparked users' emotions helped give rise to an entire industry of fake-news sites.

The apps start out with seemingly simple motivations, as entertainment that could lead to a business: Facebook is for connecting with friends and family, YouTube is for watching videos, Twitter is for sharing what's happening now, and Instagram is for sharing visual moments. And then, as they enmesh themselves in everyday life, the rewards systems of their products, fueled by the companies' own attempts to measure their success, have a deeper impact on how people behave than any branding or marketing could ever achieve. Now that the products are adopted by a critical mass of the world's internet-connected population, it becomes easier to describe them not by what they say they are, but by what they measure: Facebook is for getting likes, YouTube is for getting views, Twitter is for getting retweets, Instagram is for getting followers.

When someone goes to Google, their inbox, or their text messages, they generally know what they want to accomplish. But on social media, the average user is scrolling passively, wanting to be entertained and updated on the latest. They are therefore even more susceptible to suggestion by the companies, and by the professional users on a platform who tailor their behavior to what works well on the site.

By this point, around 2017, the public started to understand that the social media properties they loved weren't just built *for* them, but were being used to manipulate their behavior too. Spurred by public and media outcry, all of these products faced reckonings for what they'd wrought on society. Except Instagram, which largely evaded criticism. Instagram was the newest, founded four to six years later than the others, so users were still catching up to such effects, which weren't as immediately offensive and visible during the user experience on Instagram as on the other sites. Instagram, through its community and partnerships teams curating and promoting the work of its most interesting users, had done a good job of generating goodwill for its product. That work was "like making frequent deposits into a bank, waiting for an inevitable rainy day," one executive said.

But Instagram was not without problems. Its most prolific users were doing whatever they needed to do to build their brands and businesses on the site—warping reality in the process.

* * * * * * * * * * * * *

Even before the fallout over Russian influence in the U.S. election, the Federal Trade Commission had been looking into a different kind of covert manipulation, driven not by politics but by economics: influencer advertising on Instagram.

It started with a paisley-patterned dress. Lord & Taylor, a retailer, paid 50 different fashion influencers on Instagram between $1,000 and $5,000 to wear the same blue-and-orange dress on one weekend in 2015. Using captions approved by the company, the influencers had to include the hashtag #designlab and tag @lordandtaylor. But importantly, they didn't have to say they were paid.

That was a problem, the FTC said. The regulator made an example of Lord & Taylor in 2016, with a settlement saying it needed to stop its unfair and deceptive advertising practices. "Consumers have the right to know when they're looking at paid advertising," explained Jessica Rich, the director of the FTC's Bureau of Consumer Protection.

The warning shot had little impact on this thriving new economy of influence. As Instagram grew, so did the set of people willing to take money in exchange for posting about their outfits, vacations, or beauty routines, choosing their "favorite" brands with financial incentive to do so.

In March 2017, regulators sent a polite request to 90 different brands, celebrities, and influencers. The letter was intended to be a warning. Influencers now needed to let the public know when they were being paid to post something and include a disclosure at the top of the caption, not hidden in a pile of hashtags or after a long description, or else face fines. The sponsorship had to be clear and unmistakable, not something like #thankyouAdidas, or a hashtag like #sp, which some influencers were using as shorthand for "sponsored."

Once the FTC made its rules clear, Instagram built a tool that would allow brands to turn influencers' posts into actual ads, with a clear label at the top, trying to encourage disclosure. The company made it a violation of their rules to post sponsored content without using that format, seemingly taking the FTC seriously.

But then Instagram didn't enforce the policy, because after it'd made a tool that would allow users to comply, Instagram transferred any liability it might have had to the influencers and advertisers themselves. An early influencer marketing agency called MediaKix found that for the top 50 Instagram influencers, 93 percent of the posts that mentioned brands didn't adhere to FTC disclosure requirements.

A couple months later, the FTC escalated the warning, directly notifying 20 different stars and influencers, including actress and singer Vanessa Hudgens, supermodel Naomi Campbell, and actress Sofia Vergara, that they might be in violation. Campbell had posted a few suitcases from the luggage-maker Globe-Trotter for no clear reason. The singer Ciara posted baby sneakers saying, "thank you @JonBuscemi," tagging the fashion designer, without saying whether she got them for free.

· · · · · · · · · · · ·

The FTC's warnings mattered little. The company's rules mattered even less. The product's built-in incentives—likes, comments, followers—ruled all. Sponsorship deals or not, everyone on Instagram was selling in some way. They were selling an aspirational version of themselves, turning themselves into brands, benchmarking their metrics against those of their peers.

Thanks to Instagram, life had become worth marketing—not for every Instagram user, but for millions of them. Professionals, attempting to make their marketing stick, wanted to appear as #authentic as possible, as though they were tastemakers letting fans in on a secret instead of human billboards. If it worked, money from touting products wasn't the only reward. There was a chance to be one's own boss, to become an entrepreneur, or get discovered as a talent, to sell not just products, but an entire lifestyle. Influencers were thinking of Instagram not as social media, but as publishing.

"Content is a full-time job, all the time," said Lauryn Evarts Bosstick, who runs @theskinnyconfidential, an Instagram account tied to a blog, podcast, and book to share motivational messages and tips for living life well. Her account has a cohesive aesthetic: busty selfies, skintight outfits, and hot-pink visual accents. Her brand deals, usually for hair products or face cream, fit the theme. Half her revenue comes via Instagram, where her life looks like that of a Barbie doll if it were real. "I've missed birthday parties, family stuff, and people look at my account and think I'm always on a vacation," she tells me.

She started posting on Instagram while bartending in San Diego. Every day for three years, she'd use her breaks to curate the account in the bar's bathroom, until she had enough of a following that she could afford to live off her Skinny Confidential brand. Now she has almost 1 million followers. "It comes down to, how bad do you want it? You're running a full online magazine every day by yourself and you're the creative director, editor, writer, marketer, and putting it out there, hoping people like it, and then doing it all over again."

Influencers explain that Instagram provides them with immediate feedback about what people like, from all the measurable reactions—and the results aren't that surprising. Selfies perform better than photos of landscapes. Showing more skin performs better than covering up. A cohesive purpose for an account performs better than randomness. Pops of color perform better than monotone. Beauty performs better than not. Doing something visually extreme performs better. Users tweak their strategies based on the numbers, until they arrive at repeatable good results. And those good results encourage a manifestation of airbrushed selfies, crazy action shots, and scantily clad influencers.

To figure out what influencers to hire, brands would look at their engagement rates—calculated by adding likes and comments on a post, then dividing by the number of followers, using third-party services like Captiv8 and Dovetale—trying to determine whose reach was real, and whose percentage was too low to be worth the money. Like any system, it can be gamed. And Instagram ended up fueling a problem not just about truth in advertising, but about truth in life.

· · · · · · · · · · · · ● ●

The most famous instance of Instagram deception started with a bold influencer campaign and ended with a New York–based hustler sentenced to six years in prison.

It was called the Fyre Festival. The public first found out about the now-infamous spring 2017 luxury music event exclusively through Instagram, from a series of posts by some of the world's top supermodels, including Bella Hadid, Kendall Jenner, and Emily Ratajkowski. They promoted a video that showed a dreamlike experience ready to be Instagrammed: all of them together, hanging out in the Bahamas, frolicking on the beach in bikinis, dancing on yachts, and taking out Jet Skis on clear blue waters. Marketed as the party of a lifetime, the music festival was supposed to take place on a private island in the Caribbean owned by former Colombian drug kingpin Pablo Escobar, over "two transformative weekends," featuring experiences "on the boundaries of the impossible," the video promised. The festival food would come from a celebrity chef. Tickets ran up to $12,000—or more, if you wanted to reside in the $400,000 "artist's villa" and hang out with one of the performers. There were supposed to be 33 musical acts, including Blink-182, Major Lazer, Tyga, and Pusha T.

Behind the festival was marketer slash con man Billy McFarland, a mastermind of hype, who got the rapper Ja Rule to be a cofounder and the digital marketing group FuckJerry to promote it. McFarland understood the power of influencer marketing, paying Kendall Jenner $250,000 for a single Instagram post to drive ticket sales. He preyed directly on the lifestyle Instagram influencers valued. There would be exclusivity; people would get to fly in on a custom VIP Boeing 737. Guests would stay in eco-friendly luxury domes. They would be asked to preload money onto wristbands, to have a completely cashless experience. The problem was, McFarland knew more about hype than about event planning, and ended up not having anything to show for his promises.

When guests arrived, there was no private island, only a stretch of beach near a Sandals resort. There were no villas, only disaster relief tents, with their insides and bedding soaked through from the tropical rain. The cash wristbands turned out to have been a means for McFarland to get some last-minute liquidity when his project was running low on money. The most iconic image posted from the event was not of sun-kissed models or white sands, but of a sad sandwich, in a clamshell take-out box. Two slices of bread, two slices of cheese, and a side salad with dressing. It went immediately viral—on Twitter.

After an FBI investigation and a class-action lawsuit, McFarland was arrested and sentenced to six years in prison and forced to pay $26 million in restitution.

· · · · · · · · · · · • •

Most fakery on Instagram isn't subject to an explosive criminal investigation, the way McFarland's was. Instead, it goes mostly unnoticed; it's just people, behaving the way other people want them to behave, because it's a good business decision. Those living an Insta-worthy life become sources of entertainment and escapism for those who aren't.

Every day, Camille Demyttenaere and her husband, Jean Hocke, choreograph experiences entirely for the purpose of posting them on Instagram. Once, through the open door of a dark teal train curving through a jungle in Sri Lanka, Demyttenaere lunged out the side of the train into a passionate kiss with Hocke. She leaned forward, both arms fully extended behind her while hanging on to the side of the train with both hands, on top of him, with her knee up near his biceps, as he leaned out and back, his left arm dangling, holding on with only one hand, hovering over treetops.

"ONE OF OUR WILDEST KISSES," the travel influencers, who go by @backpackdiariez, said in the May 2019 caption. Commenters reacted immediately. "Are you really ready to die for a pic????????" one said. The media picked up the story internationally, writing about travel culture and the dangerous lengths people were willing to go for the 'gram. Several of them cited a study that logged 259 deaths during attempted selfies between 2011 and 2017, mostly by people in their early twenties taking unnecessary risks.

Ironically, the outrage was the best reaction the Belgian couple could have hoped for. Now, with their profile posted in more news sites, they increased their exposure and gained about 100,000 followers with a tripling of views for their Instagram stories. Their inbox overflowed with offers from tourism boards and hotels, which had discovered their profiles in the international coverage of the incident.

But everything was carefully planned, they explained. Before they go to a country, they research the best spots for photos, looking at local photographers' Instagrams and coming up with poses that haven't been done yet. (Previous influencer couples had tried the train too, without going viral.) They pick out outfits to complement the scenery. They shoot in the morning and late afternoon, when the lighting is softest. Usually they use a tripod; in the case of the train, Hocke's brother assisted, with a camera set to take 50 pictures per second. They edit their photos in Adobe Lightroom, picking the best of between 500 and 1,000 shots, removing anything unsightly like trash, shirt wrinkles, and other people, which they learned in YouTube tutorials. As a last step, they apply one of their preset Lightroom filters, which automatically tweak the shots to fit a certain mood, making the colors more saturated. They also sell the filters to the public on Instagram in packages for $25, so their audience can mimic their content if they want to.

Thousands of people travel the world to pose attractively on behalf of brands. Demyttenaere and Hocke were previously business strategy consultants in London, she for Arthur D. Little and he for McKinsey. While documenting their extended honeymoon, they attracted thousands of followers, and then realized perhaps they could extend their trip indefinitely, using their business instincts to grow it.

And it's paid off. Their clothes and sunscreen are free, as long as they mention the brands who provide it in a post. Also free are their hotels, transportation, and meals—often sponsored by a tourism board or travel agency. Brands pay travel influencers a per-post rate of about $1,000 per 100,000 followers, they said. But they make the most money off their Lightroom preset filters. Before the train incident, they were making upward of $300,000 per month just selling those via a link in their Instagram bio, Hocke said. He expects revenue to rise with follower count.

The market for travel reached $8.27 trillion in 2017, up from $6 trillion in 2006, due in part to "increased awareness among youth about travel destinations with growth in social networks," according to the World Travel and Tourism Council. This increased awareness is thanks to people like Demyttenaere and Hocke, who aren't household names, but are essentially models, paid to pose with products and encourage others to have the same adventures. They do what the feedback from their followers dictates they should do: they are photographed together, not from too far away, looking madly in love, showing glowing, tanned skin. And they feel the churn. As Hocke explained, "You need to keep feeding the machine. You always need to produce content. People think we live the life of our dreams, which is true, but you're always thinking, 'Where can I find good content, good content, good content?'" They create entertainment and escapism, like reality television with a message of bliss instead of drama, that their followers continue to like and reward.

• • • • • • • • • • • • • •

Instagram blended personal life with brand marketing at an unprecedented scale. As @instagram modeled the behavior the company wanted to see on the app, the business's ad and influencer economy supercharged the effect.

The app's users, inspired by seeing people they follow out doing interesting things, tend to want to do the same, spending their money on experiences over products. "The quest for likes requires a constant stream of new shareable content in the form of stories and pictures," the consulting firm McKinsey wrote in a report. "Experiences play into this thirst for content because they are more likely to lead to such stories and pictures than the purchase of a new product would be. Even experiences that don't turn out as expected—say, a long flight delay or rainy football game—eventually turn into shareable stories."

The Instagram effect has made it harder to sell expensive tangible products, like cars and clothing. Nine major retailers in the U.S. filed for bankruptcy in 2017, and many more closed their stores. Besides the rise of Amazon, analysts cited the experiences-not-things trend for affecting retailers' bottom lines.

Photos of leisure time are the new status symbols. People line up for hours to buy giant rainbow cotton candy at the Totti Candy Factory in Tokyo, or go to Purl bar in London for a cocktail served with a helium balloon or billowing honey fog, or pursue vacations in more picturesque settings like Iceland and Bali. In 2018, the number of plane passengers reached a record 4.5 billion, on about 45 million flights worldwide.

New businesses emerged to make it easier to get an eye-catching picture without traveling. At the Museum of Ice Cream, which started in New York in 2016 and expanded to San Francisco, Miami, and Los Angeles, visitors stand in line to be photographed while immersed in a pit of colorful sprinkles—sprinkles that are not edible, but that are made of antimicrobial plastic. At a selfie factory called Eye Candy in Toronto, for an entry fee, people can pose in one of a dozen rooms, such as one that makes it look like they're relaxing in a private jet, complete with prop champagne, and another meant to make it appear they're in Japan during cherry blossom season. At Meow Wolf in Santa Fe, New Mexico, the settings are more surreal. Billing itself as an experiential art collective, Meow Wolf invites visitors to walk through a forest of neon trees, or load themselves into a clothes dryer that appears to be a portal into another universe. And they're not slowing down; they've raised $158 million from investors in 2019 to expand across the United States.

· · · · · · · · · · · · • ·

As people curated their lives for their Instagram feeds, they also invested in enhancing their pictures, downloading apps like Facetune and Adobe Lightroom to adjust the whiteness of their teeth, the shape of their jaws, and the appearance of their waistlines. Facetune was Apple's most popular paid app of 2017, selling more than 10 million copies generally priced at $4.99.

"I don't know what real skin looks like anymore," the model and prolific internet commentator Chrissy Teigen tweeted in February 2018. "People of social media just know: IT'S FACETUNE, you're beautiful, don't compare yourself to people ok."

These editing tools made it easier for anyone who had concerns about their appearance—say, teens with acne—to continue to participate in the fun of Instagram. But they also raised the bar for what was Instagram-worthy. Dustin Hensley, a high school librarian in rural Appalachian Tennessee, said his students are only comfortable being raw and unedited on their finsta accounts, not their public Instagrams. "Anything that goes into the main account will have editing," he explained. "Generally nothing will be posted without it."

But once people advanced beyond filters to virtual nipping and tucking, seeing how much better they could look digitally, some of them started asking for those benefits in reality. The worldwide market for Botox injections to reduce the visibility of wrinkles is expected to double in size in a little over five years, reaching $7.8 billion in 2023, up from $3.8 billion in 2017. The market for synthetic skin fillers, to plump up areas with wrinkles, adjust the jawline or make lips fuller, is undergoing a similar expansion, even among teens.

Dr. Kevin Brenner, a plastic surgeon for high-end clients in Beverly Hills, has operated his private practice for 15 years, focusing on breast and nose surgeries and revisions. Dr. Brenner reports that his business has changed dramatically since the advent of Instagram. Prospective patients want to see before-and-after photos and videos of specific procedures, which he provides on his @kevinbrennermd account to 14,000 followers. Then they come in knowing exactly what they want to get done. Often, they're willing to be filmed under the knife, so that he can continue to educate his audience.

The problem: what is portrayed on Instagram isn't always feasible. He says that his competitors, the most prominent of whom have followings in the hundreds of thousands or millions, might Photoshop out a scar from a breast implant, though it's impossible to perform the procedure without making an incision. Their patients may post a before-and-after photo, where the after photo is filtered and edited and the patient's skin looks more tan and smooth than it was previously.

"A lot of times I have to manage expectations," Brenner said. "They show me a picture of someone that had something done, and they don't realize that they had it morphed through an Instagram filter." In fact, the *JAMA Facial Plastic Surgery* medical journal, published by the American Medical Association, published a 2017 article, "Selfies—Living in the Era of Filtered Photographs," which noted, "These filters and edits have become the norm, altering people's perception of beauty worldwide."

It doesn't help that in the state of California, all anyone needs to have in order to offer a plastic surgery procedure is a medical license. The American Society of Plastic Surgeons, which requires doctors to go through a plastic surgery residency to join, has a code of ethics that punishes those engaged in false advertising. But even if a doctor isn't a true plastic surgeon, they can still call themselves one on Instagram.

The most dangerous case of false expectations centers around the Brazilian butt lift. The BBL procedure was performed on more than 20,000 people in the U.S. in 2017 by board-certified surgeons, up from 8,500 in 2012, and, according to the American Society of Plastic Surgeons, was the hottest growing plastic surgery procedure in 2018. Inspired by Kim Kardashian, a BBL involves a surgeon sucking out fat from one's stomach or thighs and injecting it into their buttocks, for a body type that appeals on Instagram. The results can even be deadly if the fat cells get injected into the gluteal muscle. In 2017 a task force representing board-certified doctors found that 3 percent of surgeons performing the procedure had had a patient die.

Brenner says he doesn't offer the butt lift. Besides the safety aspects, he thinks it looks cartoonish. "It's a fad that will pass," he said. Kardashian, who sells perfume in bottles the shape of her curves, is widely rumored to have had the procedure, but once got an X-ray to prove her backside was real.

. . . .

Instagram's product always seemed to endorse an enhanced reality. The first filters, by Kevin Systrom and then Cole Rise, turned photography into art. And then, as photo-editing technology improved, models and celebrities, in their meetings with Systrom and Charles Porch, often asked for filters that would beautify their faces. With the new Stories product, Instagram obliged, building options people could try on their selfies before posting. They even let Kylie Jenner make her own filter, to allow the public to sample her lipsticks virtually.

The more people who were successful on Instagram, especially by building brands that made the human experience more visually interesting, the more successful and important Instagram became. Generally, Kevin Systrom had no qualms with the hustle, except when he did. It was tricky to determine where the company should draw the line between hustle and fraud, and its implementation of policies was inconsistent and confusing for users trying to build their businesses.

Instagram enforced against at least one aspect of fakery on its site, updating its terms of service to ban third-party services that allowed people to turn their accounts into automated like-and-comment bots, in order to get their accounts noticed by other people who might follow them. In April 2017, Instagram banned the main provider, Instagress, and it shut down. "Sad news to all of you who fell in love with Instagress: by request of Instagram we've closed our web-service that helped you so much," the company tweeted.

But it didn't change the practice, just spurred dozens of marketing blogs to write posts linking to Instagress alternatives allowing Instagram users to purchase followers and increase engagement, like Kicksta, Instazood, and AiGrow. Many are still running today.

If people couldn't pay for bots to grow their following, they weren't going to let these new rules against bots stop them. Instagram users joined pods, or groups of other like-minded Instagram users where you would quickly like and comment on the content of everyone else in the group.

"Join this Instagram pod and beat the algorithm! Share your best post here!" a group on Reddit advertised in 2019.

"If your profile about Natural/Organic Living, Tea, Herbs, Mindfulness, then drop name of your IG page. Only quality pictures, 500+ followers please," another post offered.

"Looking for members for a small but very active POD who primarily post pictures of motorcycles, or about things related to motorcycles," said a third.

Pods, which tend to be run through messaging apps like Telegram, Reddit, or Facebook, ban their members if they don't adhere to the rules about supporting each other. Some influencers even use automated services to participate in the pods on their behalf.

For those marketing on Instagram who aren't part of pods, it's hard to be seen on the feed. Edward Barnieh, the organizer of InstaMeets in Hong Kong, whose very own London couch pic promoted by Instagram's community team helped him get noticed, has seen less engagement in recent years. "My reach is crazy low and it's falling. There are a lot of people out there who don't realize that these pods exist, who think their art is bad, or their photography is worse than it is, because they're not playing the game." Instagram's solution, for anyone who asks, is to post better content—an answer that ignores how the app's system has been gamed.

．．．．．．．．．．．．．．

Businesses born of Instagram that have had better luck are those that leveraged the psychology of their users—the need for followers and recognition—while simultaneously creating interesting content. They used regular people to tell their stories, promoting them while simultaneously promoting their brands, mimicking the way the @instagram account pro-moted up-and-coming users.

Makeup brands have become especially adept. Dubai-based Huda Kattan's @hudabeauty, which has 39 million followers and sells lines of heavy, high-pigment makeup perfect for creating airbrushed looks for Instagram, constantly features videos of customers expertly applying it. Everyone whose video is chosen gets immediately thrown in front of that audience of millions. That gives anyone on Instagram hope that they may get chosen for a feature, if their video using Kattan's products is good enough. So they try, and buy more, and tell their followers to buy more. At the end of 2017, an investment by private equity valued the company at $1.2 billion.

Glossier, with 2 million Instagram fans, applies the same strategy. When Emily Weiss launched her first beauty line, she did so after years of running a blog called *Into the Gloss*, where she reviewed products and featured up-and-comers in beauty. When she launched her own brand, Glossier, *only* on Instagram, she said of it: "Who are we? We are you, listening to everyone, absorbing all of this information over the years, and trying to get at the core of what beauty is—and needs."

Glossier brought its own users into the picture, as promised. In 2016, Cecilia Gorgon, a student at the University of Michigan, applied one of Glossier's most popular products, Boy Brow, for a selfie. The company thought she looked so good wearing the product that they developed a marketing campaign around her story. "Watch out. If you tag Glossier in a selfie you could end up like this," the company told its followers.

In 2018, this Instagram-first company passed $100 million in annual revenue and acquired 1 million new customers, all through direct sales. That year, Glossier sold one Boy Brow every two seconds. Their few retail locations are designed to offer experiences and function as market-ing venues more than as sales outlets. In the Los Angeles location, there is a mirror with the words "You Look Good" inscribed so it will show up in a photo; everything is painted millennial pink; all makeup can be tried on on the spot; and the lighting is specifically designed for phone photography.

In the back of the showroom, one of the building's closets has been transformed into an immersive replica of the picturesque rock formations in Antelope Canyon, so visitors to the Glossier store can pretend they were actually in the photogenic natural landmark. Glossier plays sound recorded in the actual canyon, so that it works for video too.

• • • • • • • • • • • •

All of this perfection and commercial work masquerading as regular content has a price: a feeling of inadequacy for users who don't understand the mechanics behind the scenes.

In May 2017, in a widely publicized study, the Royal Society for Public Health in the U.K. named Instagram the number one worst app for mental health for youth, specifically because it drives people to compare themselves to one another and fosters anxiety. "Seeing friends constantly on holiday or enjoying nights out can make young people feel like they are missing out while others enjoy life," the report said. "These feelings can promote a 'compare and despair' attitude in young people. Individuals may view heavily photo-shopped, edited or staged photographs and videos and compare them to their seemingly mundane lives."

The RSPH looked at all the big social platforms, including Snapchat, YouTube, Facebook, and Twitter, and made recommendations. Ideally, they said, apps would let users know if they were spending an unhealthy amount of time glued to the screen, or if they were viewing medical information from a valid source. They suggested schools teach tactics for social media health, as seven in ten young people have experienced some form of cyberbullying. Some recommendations targeted Instagram specifically—for example, the suggestion that apps indicate when a photo or video has been edited, perhaps "in the form of a small icon or watermark at the bottom of someone's photo that indicates an airbrush or filter has been used that may have significantly altered their appearance." On Instagram, users are so accustomed to enhanced images that the culture of disclosure works the opposite way, with people tagging photos #nofilter when it's real.

Instagram's work to introduce the Stories product and reduce the pressure on the app increased the amount people were willing to use Instagram, solving its growth problems. But it did not change the underlying culture of the app.

That was what Systrom's well-being initiative was supposed to accomplish. It was supposed to launch Instagram into some historic echelon of positive innovation, creating a ripple effect of healthy changes around the rest of the internet. But for months, Instagram couldn't get out of its own way to build anything beyond a comments filter.

Even for the team to define "well-being" was a monstrous task. Nicky Jackson Colaço, the head of policy, thought it couldn't be as simple as banning more things. Facebook had been pretty much in charge of enforcing Instagram's content rules—the ones against nudity, terrorism, and violence—ever since the acquisition, and not doing a good job. Jackson Colaço decided the well-being initiative should go beyond that, improving users' experience on Instagram more broadly to make people happier and healthier.

But every time her team presented to Systrom what the specifics of the plan might look like, he would explain that it didn't look quite right to him, and that they should keep thinking. Jackson Colaço worried that if the team didn't apply specifics to the plan soon, it would end up as a pure marketing campaign, not the visionary idea she thought could win Systrom awards and an honored place in internet history. In reality, Systrom was in a position with Facebook where every move was scrutinized. He needed to pick his battles for resources.

Instagram was only one step ahead of the public. As the well-being group presented to the rest of Instagram employees at one of their Friday meetings, touting and celebrating product ideas that were nowhere near production, and as Instagram's leadership deliberated about what to build and what to call it, the wider world started to catch on to the app's downsides.

Instead of resolving the many debates around the well-being initiative and settling on a broader strategy, Systrom pushed deeper on the work that was already getting Instagram praise, which could use some of Facebook's resources: comment filtering.

Technologically, Systrom decided to build upon an artificial intelligence tool from Facebook. The machine learning software could learn, over time, what was contained in a post, in order to classify it and provide better intelligence to Facebook about what people were sharing. Systrom thought it would be interesting to apply the same technology to user comments, to try to identify and block the unkind ones. A group of employees sorted through samples of Instagram users' comments, rating them on a scale of 0 to 1, doing work the machine would eventually take over.

On the community side, Instagram ran a #kindcomments campaign, with celebrities like actress Jessica Alba and plus-size model Candice Huffine reading the most inspirational responses on Instagram. They enlisted artists to create murals around the world, from Jakarta to Mumbai to Mexico City, celebrating kindness and lifting up others.

Instagram's new comment filter would take all the worst comments and just make them not exist. Few users would notice the new default setting on their apps. It would just make Instagram seem more pleasant than it actually was.

But pushing any further than this, into the bigger issues Instagram was facing, was a matter of priorities. Instagram didn't want to spend all its time on cleanup. The team was inspired by its ability to draw in more people with Instagram Stories, and wanted to continue to prove that it could make new things people would enjoy using. It would be a few more years before Instagram took its next big swing at addressing feelings of inadequacy on the site, with a test removing like counts in 2019.

• • • • • • • • • • • •

Facebook's culture for responding to crisis was fully reactive: the company addressed problems only once they resulted in major blowups that politicians and the media were paying attention to. With the crisis over Russian influence, Facebook was in a frenzy, while Instagram was insulated. At the time, Instagram only sent a couple communications and policy employees part-time to Facebook's internal war room, so they could help figure out what had happened and answer government questions.

But the rest of the team was updating the product as usual, improving Instagram Stories and the new algorithm.

Later most of them would be disturbed to read, in December 2018, after a couple years of feeling superior, that Instagram was actually not so innocent. That month, research groups commissioned by the Senate Intelligence Committee would report that the Russian Internet Research Agency (IRA), the troll farm that had run the campaign to divide America with memes and fake accounts, received more likes and comments on their Instagram content than on any other social network—including Facebook. While Facebook was a better venue for going viral, Instagram was a better place to spread lies.

On Instagram, anyone could become famous among strangers. And so the Kremlin's IRA did too. Nearly half of their accounts achieved more than 10,000 followers, and 12 of them had over 100,000. They used the accounts to sell things. One sold the idea that Hillary Clinton was a bad feminist. Another, @blackstagram_, with 303,663 followers before Facebook took it down in the Russian account purge, touted products from what it said were black-owned businesses, while telling black Americans not to waste their time voting.

When the Senate committee posted the report stating that Instagram was just as much a hotbed of Russian misinformation as the rest of the internet, the media spent a day writing about it, and then moved on. The Senate asked for no extra testimony. People liked using Instagram. They went back to talking about Facebook, and holding Facebook accountable for its wrongdoings, not acknowledging that the two were one and the same.

Perhaps assigning blame to Facebook was appropriate. Facebook, after all, wanted the credit for Instagram's success. But during the tussle for power in 2018, leaders at both social networks would fail to prioritize fixing Instagram's downsides.

THE CEO

"Everything breaks at a billion."

—FORMER INSTAGRAM EXECUTIVE

T he debate about whether Instagram threatened Facebook's dominance was starting to color every interaction between the two leadership teams, especially when it came to hiring. Instagram couldn't just go out and pick the people they needed to get something done. Kevin Systrom and Mike Krieger had to make a detailed pitch to Mark Zuckerberg, and only he could decide if the head count was worth it. Every Facebook team had to do this, but not every team at Facebook was running a mini company within the overall company, with its own revenue and product that didn't depend on the Facebook news feed.

Zuckerberg told Instagram they could hire 68 people in 2018, increasing their workforce by about 8 percent. For the founders, that number was shockingly low. They had plans to invest in addressing Instagram's problems, as well as to develop a bold video section of the app called IGTV, which they hoped would be as well received as Stories had been. Meanwhile, employees were having trouble supporting the growing network's needs.

They needed to fight back with data. Krieger put together a chart for Zuckerberg that compared Facebook and Instagram based on employees per user. In 2009, when Facebook had 300 million users, it had 1,200 employees. In 2012, when Facebook reached 1 billion users, it had 4,600 employees. Instagram was probably going to reach 1 billion users in 2018, but it had fewer than 800 employees. They weren't adding people nearly as quickly as the app was growing.

But Zuckerberg wasn't as moved by data in this instance as he usually was, because he didn't imagine Instagram being so independent in the future. Now that he knew every Instagram success might result in a blow to the longevity of the main social network, it was more important to him than ever to coordinate between the teams. Facebook and its employees—and Zuckerberg himself—would have to be more directly involved in whatever Instagram did next, removing some need for hires.

Zuckerberg told Krieger and Systrom they could add 93 more people. It was better than 68, so the founders felt a little victorious—until they found out how many new employees were going to other, less lucrative segments of Facebook. Facebook Inc., with more than 2 billion users of its main social network, would end up adding 8,000 employees around the world in 2018, to reach a total of more than 35,000.

"How much head count did Oculus get?" Systrom asked Brendan Iribe, the cofounder who was no longer CEO of the virtual reality division that Facebook had acquired for $2.2 billion in 2014 but was still working there.

"More than 600," Iribe said.

Instagram was on track to deliver $10 billion in revenue in 2018, while Oculus was on track to lose millions. They were very different types of businesses, but still Iribe agreed it was unfair. In that moment, Systrom realized that all the work Instagram had done—building the second-biggest social network, developing the first significant line of revenue since news feed ads, helping draw the attention of young people and celebrities, evolving the culture of the world—would not be rewarded with the support it needed to keep making significant strides.

It wasn't just Oculus. More comparable segments of Facebook, like the video initiative to compete with YouTube, were getting to hire hundreds of people too. So at Instagram, one of the fastest-growing parts of the business, on track to produce 30 percent of Facebook's revenue by 2019, resentment and frustration started to brew.

· · · · · · · · · · · ●

To an outsider, Instagram's branding still appeared quite independent. Nobody was talking about election interference or fake news on Instagram. Besides its ultra-Instagrammable headquarters in Menlo Park, Facebook was close to leasing space in San Francisco and New York for the app, where they'd build even more interactive and visually interesting offices, perfect for hosting celebrities. But internally, the relationship between Instagram and Facebook was getting more political than ever.

Krieger and Systrom had always joked that their partnership was so harmonious because neither coveted the other's job. Krieger didn't need to be the product-obsessed face of the company, and Systrom didn't need to be the behind-the-scenes architect. In December 2017, they got to test their theory. Systrom and his wife, Nicole Schuetz, had their first child. So for about a month, Krieger took on Instagram CEO duties, and the experience confirmed for him that what they'd been saying was true: he would never want Systrom's job, or at least not what it had become.

So much of the role was about negotiations with Facebook. Over the winter of 2018, the debate concerned Instagram's plan to launch its IGTV app, devoted to longer-form video, in a vertical format so people wouldn't have to tilt their phones to watch. Instead of Instagram just going for it after giving Facebook the courtesy of a heads-up, Krieger, usually deep in planning for engineering and infrastructure or helping employees understand Instagram's product philosophy, was spending his time dealing with bureaucracy, shuttling back and forth for meetings at Facebook's offices in Building 20, with Zuckerberg, Facebook chief product officer Chris Cox, and Fidji Simo, the head of video, who was in charge of Facebook Watch.

Zuckerberg thought that with IGTV, it was Instagram's turn to help Facebook grow. Facebook Watch hadn't caught on with users, even though Facebook had put substantial resources behind it, paying studios and news organizations to create shows for it. He wanted IGTV to be built in such a way that it could integrate with Watch and feed content there, so Simo came up with a presentation about the ways it could work.

Krieger had always told Instagram employees to "do the simple thing first." He thought it didn't make sense to have all these discussions that would only be relevant if the product succeeded. If IGTV became popular, *then* they could talk about helping Facebook. "We'd be lucky to have that problem," he'd say.

When Krieger finally got approval to build a separate app, after a month-and-a-half delay, Zuckerberg dropped another bombshell: everyone was going to get a new boss.

• • • • • • • • • • • •

The new hierarchy, the biggest reshuffling at the top of Facebook in corporate history, would formalize the new way Zuckerberg thought about his acquired properties Instagram and WhatsApp. Those two apps would be bundled with Facebook Messenger and Facebook itself, as part of a "family of apps," all reporting up to Chris Cox, who was Zuckerberg's most trusted product executive.

Zuckerberg wanted to create more navigation between the apps, so that their users could switch between them easily. He gave the integrations a friendly term: "family bridges."

A lot of employees were skeptical as to whether the public *wanted* bridges between the apps, since people used them for different reasons. After the U.S. election and all the privacy debacles, the public was still wary of Facebook in a way they weren't yet of Instagram or WhatsApp. But Zuckerberg's word was final. He was working off data that proved that more connections rendered a network exponentially more useful. He chose to prioritize that data, as opposed to the data that showed people in larger networks share less. If it all worked smoothly, Zuckerberg would be able to create the ultimate social network. Facebook would be as big and powerful as the "family" was.

As with most families, there was drama. The parent company was still, in the eyes of regulators, in trouble for not being transparent about Russia's interference in the U.S. election. In the first quarter of 2017, right after the election, Facebook started a "lockdown" so a group of employees could build tools to prevent people with false identities from manipulating future elections around the world. The tools enabled Facebook to catch some similar efforts but were far from foolproof.

Instagram was an afterthought to that conversation. WhatsApp too was relatively uninvolved in the election discussions. Zuckerberg considered that the apps could be a hedge against Facebook's issues, by providing more surfaces for advertising and more ways to draw people into the overall network. For that reason, Systrom and Krieger's relationship with Zuckerberg looked calm and peaceful in contrast to that of the WhatsApp founders. Instagram was helpful to Facebook's business, but in early 2018, the messaging app, purchased for $22 billion, had 1.5 billion users, and still no clear path to making money.

Facebook pushed to put advertising in WhatsApp Status, their version of Stories. But in order to place those ads in front of the right people, WhatsApp would have to know more about the users of the chat app, which would mean chipping away at the encryption. The founders, Brian Acton and Jan Koum, stubbornly resisted the idea, which violated their motto—"No ads, no games, no gimmicks"—and which they thought would break users' trust.

Acton decided to leave Facebook: his decision cost him $850 million in stock options. (He was still a billionaire from the deal, many times over.) Koum, WhatsApp's CEO, planned to leave the company that summer. Later Acton told *Forbes* reporter Parmy Olson that Facebook "isn't the bad guy. I think of them as just very good businesspeople."

Whatever they wanted to do was their right, he said. He could choose not to be a part of it, but he couldn't prevent it from happening. "At the end of the day, I sold my company," Acton underscored. "I sold my users' privacy to a larger benefit. I made a choice and a compromise. And I live with that every day."

．．．．．．．．．．．．．．

Facebook executives talked among themselves about how ungrateful the WhatsApp founders had been. The consensus was that the team had always been high maintenance, asking for slightly bigger desks, longer bathroom doors that reached all the way to the floor, and conference rooms that were off-limits to other Facebook employees. If they wanted to leave in a huff at the slightest suggestion of making the investment worth it, after Zuckerberg had made them both billionaires, then good riddance. "I find attacking the people and company that made you a billionaire, and went to an unprecedented extent to shield and accommodate you for years, low-class," David Marcus, the Facebook executive in charge of a new cryptocurrency initiative, later wrote publicly. "It's actually a whole new standard of low-class."

It showed what could happen if, as an acquired company, you didn't realize you were still beholden to Facebook's needs. But Systrom and Krieger felt like they'd been much more reasonable. Besides their ad business, they'd suffered all those IGTV meetings and talks of "cannibalization." They'd begrudgingly built more prominent ways to navigate to Facebook from Instagram. And yet, if things continued trending in this direction, Instagram would get less independent. Painful though it was to consider, they might be the next out the door.

Having a new boss at least meant there was an opportunity to air frustrations. Though Cox had been one of the executives interested in understanding cannibalization, his tone changed when he became Instagram's boss.

"Let's be straight with each other," Systrom told Cox, with Krieger in the room, once he was back from paternity leave. "I need independence. I need resources. And when something happens, I know I'm not always going to agree with it, but I need honesty. That's what's going to keep me here."

Cox said that he was committed to advocating for everyone under his leadership, including Systrom and the new leaders of WhatsApp, Messenger, and Facebook, to have the creative freedom they needed to do a good job. That year, he decided that keeping Systrom and Krieger from leaving would be his top priority.

• • • • • • • • • • • • •

Then, as often happens with Facebook, a revelation in the media shifted everyone's priorities.

On Friday, March 17, 2018, the *New York Times* and the *Observer* simultaneously broke the news that years earlier, Facebook had allowed the developer of a personality quiz app to obtain data on tens of millions of users, which that developer then shared with a firm called Cambridge Analytica.

Cambridge Analytica retained the data and used it to help build its political consultancy. The company aggregated information from several sources to build personality profiles on people who might be receptive to ads that would help conservatives win elections. Donald Trump's campaign was a client.

The story hit all of Facebook's weak spots: shoddy data practices. Negligence. Lack of transparency with users. And a role in Trump's win. It stoked distrust among politicians the world over.

The worst part was that Facebook had known of the data leak for years, and hadn't properly enforced its policies, or let users know when their information was compromised. The company had even sent threatening legal notices to the media to keep the story from coming out. And then, for several days, as public anger stirred and Facebook users clamored to know whether their data was involved in the leak, Zuckerberg and Sandberg remained silent, fretting about what to do.

As regulators in the U.S. and Europe said they would probe the matter, Facebook stock fell about 9 percent, erasing $50 billion in market value, in just three days following the news breaking. The hashtag #deletefacebook started trending. Even WhatsApp cofounder Acton tweeted it, before deleting his Facebook account.

A week later, Zuckerberg agreed to testify in front of U.S. Congress for the first time, on April 10 and 11, 2018, under interrogation by the Senate and then the House of Representatives. The questions weren't about Cambridge Analytica as much as they were about Facebook's power. Legislators were waking up to the fact that one company, in charge of entertaining and informing more than 2 billion people, was more influential in many ways than the government itself. Things that Facebook had done for years suddenly looked scandalous under this lens.

Facebook's business model, which required collecting all types of user data on websites and apps even beyond those owned by Facebook, seemed riskier now that legislators knew that data could be leaked.

Facebook's core news feed product, where news and information could so specifically be targeted to users' interests, also seemed to have a tremendous downside. You couldn't know what someone else saw when logged onto Facebook, what shaped their reality. Some people were selling illegal drugs; some people were getting radicalized by the Islamic State; some people were not people at all, but bots trying to manipulate public conversations. Only Facebook had the power to understand and police it all—and they weren't.

And yet, there was not much legislators could do.

For one thing, they couldn't agree on the main trait they hated about Facebook. They all had different agendas to pursue with Zuckerberg. And two, some of their critiques fell flat because they didn't know nearly enough about how Facebook worked.

For example, Senator Orrin Hatch asked, "How do you sustain a business model in which users don't pay for your service?"

"Senator, we run ads." Zuckerberg smirked. The line ended up on T-shirts.

This was his disposition throughout. Zuckerberg had been trained by his lawyers to keep the testimony as dry as he possibly could. And it worked—he returned to the headquarters victorious. At least he hadn't created new problems for the company. Some employees even toasted him with champagne.

· · · · · · · · · · · ·

Congress hadn't rattled Facebook's leaders, but the Cambridge Analytica stock drop had. They started to acknowledge that perhaps their utilitarian strategy and rapid product development had led to massive blind spots in the organization. The company embarked on an audit of everything in production, attempting to see if there were unexpected flaws that if unchecked could result in scandal.

As part of the response, Facebook committed to building out its "integrity" team to be almost the size of Instagram, charged with handling all the content and privacy problems in the Facebook "family."

That team, led by Guy Rosen, strangely reported up to Javier Olivan, the VP of growth, who reported to Zuckerberg. The people thinking about fixing the product had an incentive not to fix it too much, at least not to the point where it would jeopardize Facebook's business.

Still, it was a step in the right direction. Systrom asked Rosen if it would be okay to have a portion of the new integrity hires focus on Instagram-specific issues—especially since Instagram wasn't getting many new employees, total. He worried that Instagram might find itself in Facebook's unenviable position if they didn't pay more attention to their own problems immediately, some of which were similar to what Facebook was seeing, and some of which were unique to Instagram. While Facebook was a place for people to use their real identities, Instagram users could be anonymous. While Facebook was a place where content went viral, the dangerous communities on Instagram were harder to find, discoverable only if you knew the right hashtag. Instagram wouldn't be able to catch all the worst posts on the platform simply by adopting the same policing tactics as Facebook.

Because Instagram had shifted its content moderation to Facebook after the acquisition, Systrom was disconnected from how specific issues were handled, except when it came to the company's most high-profile users. It wasn't like the early days, when they'd had actual employees going through all of the most upsetting content on the app. For the past few years, the Instagram full-time employees had been focused on shaping the community through promoting good behavior, and not paying as much attention to stopping the bad.

Instagram had a separate set of community guidelines, more tuned to a visual network where people built personal businesses. It told users not to spam each other for commercial purposes, or steal each other's content, or post unclothed pictures of their children. Everything that users reported to Instagram went into the same queue as Facebook's flagged content. Then an army of outside contractors at firms like Cognizant and Accenture sorted through it all quickly, making yes-or-no decisions on imagery that was often traumatic and scarring. The system was necessary because Facebook was a business, trying to spend as little on costly human moderation as possible. The average Facebook employee made more than six figures, while a contractor in Phoenix might make $28,800 per year, and in Hyderabad, India, $1,401 a year. Some of them lasted mere days or months, because of the burden on their mental health from seeing the worst of humanity daily.

The actual full-time employees on Rosen's team, tasked with thinking about the issues at a more systemic level, prioritized what got Facebook in the most trouble with governments around the world, like election misinformation and terrorist recruitment. Instagram seemed to be the least of their worries, because of the relative lack of scandal.

But Systrom worried about areas like live video. Facebook was investing heavily in finding live-streamed violence quickly. *BuzzFeed News* tallied that between December 2015 and June 2017, at least 45 violent acts, including murder, child abuse, and shootings, were broadcast live on Facebook. Instagram introduced live video a year after Facebook, in 2016. Systrom argued that it would make sense to staff up Instagram's defenses against live violence. The press hadn't written about any similar problems on Instagram, but Instagram's live product was much more popular than Facebook's, so it was only a matter of time before the same problems cropped up.

Because Instagram was about photos and didn't require people to use real names, it was easier to sell drugs, or advocate for suicide, or post hateful and racist content on the platform. Terrorists were recruiting on Instagram as they were on Facebook, just more covertly organizing their content with hashtags, using memes to recruit youth. Instagram needed more people focused on monitoring those things who knew how to navigate the platform, Systrom urged.

Rosen was receptive to the message. Around the same time, he was realizing the depth of some of the Instagram-specific issues, thanks in part to a message he had just received from a woman named Eileen Carey.

Carey had been trying to get Instagram's attention on the issue of opioid sales on its app ever since 2013, when she worked at a consultancy on behalf of Purdue Pharma, the maker of OxyContin. Real and counterfeit pills, as well as other drugs, were posted on Instagram openly for sale, easily rendered searchable via hashtags like #opioids or #cocaine. The posts usually included a phone number for coordinating a handoff and payment in WhatsApp or another encrypted chat app.

Every time Carey opened Instagram, she would spend a few minutes scrolling for drug content to report it to the company. She'd usually receive messages back saying the posts didn't violate Facebook's "community standards." After hundreds of reports over the years, she got angrier about the problem, even when it was no longer her job to be. Deaths from opioids had more than doubled in the United States, to 47,000 in 2017, and she thought Instagram was a key way the drugs were reaching young people. She started building a file of screenshots of the posts she'd reported, with the corresponding inaction from Facebook, to give to the media and Facebook executives.

She finally reached Rosen in April 2018 via Twitter direct message, and told him to do a search for #oxys on Instagram. There were 43,000 results at the time. "Yikes," Rosen responded. "This is SUPER helpful." For years, Instagram simply hadn't been proactively looking for drug sellers, or paying attention to the trend in Carey's reports. With Rosen's push, the company removed the hashtags one day before Zuckerberg testified.

But removing an offending hashtag or two didn't remove drug sales from Instagram. The general problem remained. So Rosen told Systrom his request for his own integrity staff made a lot of sense.

Then Zuckerberg denied it. He said Instagram had to figure out its problems with its own resources. Facebook was in trouble, so Facebook was the priority. It would be up to Systrom to negotiate for parts of Rosen's Facebook-focused team to take a look at Instagram-specific issues, if they had time.

Yet again, Zuckerberg was prioritizing Facebook's needs over Instagram's. The logic was about centralizing all the work to make it more efficient. But in practice, Systrom saw corporate hierarchies getting in the way of users' safety on the app.

● ● ● ● ● ● ● ● ● ● ● ● ●

Systrom and Krieger realized that maybe they would need a new approach in arguing for their resources and independence, especially given the public scrutiny Facebook was under. Perhaps a Facebook native could help, strategically. They had their eye on one of Facebook's top leaders, Adam Mosseri, who had been at the company for almost a decade. With a background as a designer, he ran Facebook's news feed. He had been waging battles to improve the product's look and feel. He also happened to be good at Instagram, with an account full of aesthetically pleasing cityscapes and nature shots.

They needed someone new on product. Kevin Weil, whom Systrom recruited in from Twitter in 2016, had left to join Facebook's new cryptocurrency group, Libra, which would try to develop a global form of money to rival the U.S. dollar. So Systrom and Krieger recruited Mosseri to replace Weil.

Instagram employees were skeptical of their choice and wondered if the Instagram cofounders had had a choice at all. Amid the tension with Facebook, nobody was sure if the founders really wanted Mosseri, or if they'd been forced to bring him in so that Instagram could be more tightly controlled by Facebook.

Even Mosseri was surprised to be a candidate. Mosseri had always liked and respected the Instagram founders, but he thought, over the past year, that he'd been a pain. They had had a few uncomfortable debates about small details. At one point, Mosseri had removed one of Instagram's promotions on the Facebook site, telling them they needed to redesign it if they wanted it back. Everything related to the debate about how much Facebook was helping Instagram, or how much Instagram should help Facebook, was sensitive.

Mosseri, at 35, just one year older than Systrom, was a tall, broad-shouldered, square-jawed man with curved dark eyebrows, a friendly smile, and a widow's peak hairline. Sometimes he wore hipster glasses that made his eyes look larger and more sincere. He was well-liked both inside and outside the company, which made him especially valuable to Facebook, as the social network was losing the public's trust.

Recently, in addition to doing his day job, Mosseri had become something of a company spokesman. In response to journalists' critiques, he provided Facebook's perspective, which the company would post on Twitter as part of a program to improve relationships with the media on the site the journalists used most. He met with members of Congress to explain the news feed. Four days after he started the process to interview for the Instagram job, he went on an exhausting multi-country European tour to talk to policymakers about data privacy.

With a high-profile role at an embattled Facebook, in charge of about 800 employees, Mosseri was getting a little burned out.

Mosseri had two little boys—a baby and a toddler—in San Francisco with his wife, and he was worried about being present for them. Maybe, he thought, the Instagram job would be less intense. Instagram seemed beloved by users and even the public and media, or at least much more so than Facebook. While the app certainly had its problems, Instagram had spent years telling its story, through promoting its own users and partnering with celebrities. It had successfully convinced the wider world—and coworkers at Facebook—that it was a special place for beautiful things.

Being at the top of Instagram even looked fun. The April of Facebook's doom, as Zuckerberg was followed by camera crews on his way to provide congressional testimony, Systrom passed a test to become a wine sommelier. He donned a tux to sit with the Kardashians at Anna Wintour's Met Gala, the most exclusive party in New York. While Mosseri was thinking about European data laws, Systrom was thinking about IGTV.

Mosseri didn't know the other reason the Instagram founders had recruited him. Systrom and Krieger thought that if tensions with Zuckerberg mounted, or if they became tired of navigating politics with Facebook, they needed to train someone who they could trust, who could advocate for Instagram with Facebook. One day, Mosseri might need to lead the company they'd started.

• • • • • • • • • • • • •

On a Tuesday in June, Instagram finally reached the milestone they'd been working toward: 1 billion users. This was the pinnacle they'd realized it was possible to reach after launching Stories. It was also the same metric Facebook hit the week Instagram had joined the company in 2012. Now Facebook and Instagram were peers in shaping the world through their products at a massive scale.

They reached the billion mark just in time for their flashiest product launch ever. After fighting so hard for the right to launch IGTV as a separate app without direct tie-ins to Facebook Watch, Instagram was pulling out all of the stops to highlight the differences between its vision and Facebook's.

The events team took over the former Fillmore West concert hall in San Francisco, putting a giant balloon arc over the entryway to brighten up a street otherwise dotted with homeless encampments. The launch event was meant to celebrate the photographable, filmable moments and products that had become popular on the app. Staff handed out cruffins (croissants shaped like muffins) with raspberry cream filling to press and influencers waiting in line to enter. Once guests walked up the painted steps, a wide range of fancy, colorful foods awaited them, from avo-cado toast to açai bowls guests could top with fresh berries and coconut. Nearby, a barista made matcha lattes, with many areas throughout de-signed for selfie-taking.

Instagram was putting on a show to build hype for the product and make the event itself Instagrammable. Lele Pons, the former Vine star who now had 25 million followers on Instagram, was there. So was Ninja, the famous video game streamer, and beauty vlogger Manny Gutierrez.

One thing was missing: the final copy of Systrom's presentation. Somehow, nobody could find the polished video with firework effects, formatted to fit the venue's screen, used in many smooth rehearsals. Staff delayed the event as a design executive worked to cobble together a new presentation from one of the drafts, while a couple hundred guests were already seated facing the stage, bathed in red mood lighting, waiting for something to happen.

And then something did. All the details of the product appeared in a blog post on Instagram's website. The post was timed to go up while Sys-trom was presenting, but he wasn't talking yet, and nobody had thought to reset the timer. The press wrote and published their first stories based on the blog post, while still sitting in their seats waiting for Systrom. Finally, he appeared onstage, masking his frustrations with good humor. He gave a shortened version of his planned presentation, followed by a press conference.

It wasn't pretty, it wasn't Instagrammy, but it was done. IGTV, the most ambitious thing Instagram had tried to do since Stories, was live.

All the guests and employees, after the most Instagrammable morning, were reminded of the app's owner the second they walked into the San Francisco transit station next to the venue. There, an entire hallway was plastered with posters from Facebook's expensive global mea culpa campaign. "False news is not your friend," one advertisement said. "Clickbait is not your friend." "Fake accounts are not your friends."

An hour after presenting, Systrom was reminded of Facebook too. His iPhone flashed with his new boss's name, and he went to a quiet spot to take the call. *Good,* he thought. Even if Cox and Zuckerberg weren't attending the event, they were at least acknowledging the accomplishment. He swiped to answer.

"We have a problem," Cox said. "Mark's very angry about your icon."

"Are you serious? What's wrong?" Systrom asked.

"It looks too much like the icon for Facebook Messenger."

The IGTV logo had a sideways lightning bolt shape inside a TV-shaped box. The Messenger logo had a similar bolt, but inside a cartoon dialogue balloon.

After the drama of the day, there was no praise from on high—only Zuckerberg's concern that Instagram would step on Facebook's branding.

• • • • • • • • • • • •

A month later, Zuckerberg touted both the new IGTV product and the fact that Instagram had reached 1 billion users during Facebook's earnings conference call with Wall Street investors.

For as long as Instagram was part of Facebook, it would be fair for Facebook to take some credit for their milestones. Now Zuckerberg let the public know exactly how much he thought they deserved.

"We believe Instagram has been able to use Facebook's infrastructure to grow more than twice as quickly as it would have on its own," he said. The 1 billion user milestone was "a moment to reflect on how this acquisition has been an amazing success" not just for Instagram, but for "all the teams across our company that have contributed."

Systrom had tried to get Instagram's successes into Facebook earnings calls over the years, but had rarely succeeded. Now Instagram was a star of Facebook's business plan, but in a way that gave Facebook heavy credit. Instagrammers especially noticed the "more than twice" statistic. There was no objective way to calculate that number.

The IGTV launch was Systrom's way of asserting Instagram's place in the public eye. The earnings call was Zuckerberg's. Zuckerberg needed to show to Wall Street, and the public, that Facebook was still an innovative, creative company that had many ways to grow, despite its scandalous setbacks. Zuckerberg cared so much about appearing innovative, Facebook employees regularly polled the public regarding their impressions of his leadership.

Either way, after the earnings call, Systrom let his frustrations show with employees for the first time. He and Krieger told Instagram staff that they thought the app *would* have been able to get to a billion users on their own. It might have taken longer, but maybe not twice as long.

Every time Instagram had a slice of success, Zuckerberg seemed to kick them back in their place. And it was about to get worse.

• • • • • • • • • • • • •

All the top Facebook executives came together for a routine meeting, which would take place in a conference room over the course of three days, to plan the second half of the year. Though the first half of 2018 had been marked by public criticism, the most heated debate among the internal team was about something else: Zuckerberg's "family of apps" plan.

Cox told Zuckerberg he needed to let the products build independently and not become too similar. "They'll compete a bit with each other, but if we have more unique brands, we'll be able to reach different kinds of users."

He and Systrom had spoken extensively about using Harvard professor Clayton Christensen's "jobs to be done" theory of product development, which states that consumers "hire" a product to do a certain task, and that its builders should be thinking about that clear purpose when they build. Facebook was for text, news, and links, for example, and Instagram was for posting visual moments and following interests.

But that wasn't how Zuckerberg thought of it.

"We should think of this globally," Zuckerberg said. "We're trying to build a global community—not a bunch of smaller communities." If you added up all the unique people using at least one of the apps in the family, you got a community of 2.5 billion, which was bigger than Facebook.

"I just think it's going to be hard to do," Cox protested. "These are pretty different teams, and their user bases are differentiated a lot already."

"Aren't we taking an operational risk?" Systrom added. "If I have to worry about Instagram, but also have to worry about Messenger and Facebook, I'm not sure what that looks like in practice."

"I think it's a risk we should take," Zuckerberg declared.

Other reasons existed, besides making the network really, really big. The company could present a united front to regulators on their data policy. They could have the same content rules for users of every Facebook-owned app. Theoretically, it could make Facebook more difficult for the government to break up into pieces, should an antitrust challenge arise, though that wasn't Zuckerberg's stated strategy.

Either way, the debate mattered little. Zuckerberg had already made up his mind.

• • • • • • • • • • • •

In Zuckerberg's mega-network plan, Instagram was supposed to focus on finding users that were different from Facebook's. And now that Instagram was growing its revenue and number of users faster than Facebook, he decided it was time to take the training wheels off completely. So that summer, Zuckerberg directed Javier Olivan, Facebook's head of growth, to draw up a list of all the ways Instagram was supported by the Facebook app. And then he ordered the supporting tools turned off.

Systrom again felt punished for Instagram's success.

Instagram was also no longer allowed to run free promotions within the Facebook news feed—the ones that told people to download the app because their Facebook friends were already there. That had always brought a steady stream of new users to Instagram.

Another of the new changes would actually mislead Facebook users in an attempt to prevent them from leaving for Instagram. In the past, every time an Instagram user posted with the option to share on Facebook, the photo on Facebook said it came from Instagram, with a link back to the app. Instagram's analysis showed that between 6 and 8 percent of all original content on Facebook was cross-posted from Instagram. Often, the attribution would be a cue for people to comment on the photo where it was originally posted. But with the change mandated by the growth team, that attribution would disappear, and the photo would seem as if it had been posted to Facebook directly. So there was no longer a link to Instagram on tens of billions of Facebook photos every day.

Without Facebook's help, Instagram's growth slowed to a halt. That bolstered Zuckerberg's argument that Facebook had helped them grow faster.

Systrom had never been one to criticize Zuckerberg in front of his employees. But this time he wrote a long internal message, saying that he disagreed entirely with the new strategy. Still, he said, Instagram would have to comply with the order, even if it was wrong.

After all the hours Systrom had spent in leadership coaching over the years, all the books he'd read about how to be a better CEO, and all his personal improvement quests, he was faced with an unexpected personal discovery: he wasn't the boss. He started telling his close confidants that if Zuckerberg wanted to run Instagram like a department of Facebook, maybe it was time to let him. Maybe there wasn't room for another CEO.

In need of some time to contemplate, Systrom took the second half of his paternity leave that July. And Instagram's growth team went into an immediate Facebook-style lockdown.

Lockdowns were usually instituted at Facebook for time-sensitive issues, like developing a product to beat a competitor, or working on election interference. The work hours were longer, the employee shuttles ran later, and everything else on the road map was put on hold.

This lockdown was different. The Instagram team was trying to figure out how it would grow without Facebook's help. They considered the possibility that Zuckerberg would make even harsher moves in the future, like cutting off Instagram from access to information about who was Facebook friends with whom—data that helped Instagram show people content from their closest friends.

At the end of the month, Instagram's growth team had made enough changes to the app to reverse the slowdown, exceeding their goals. In a way, recovering was easier than they had expected. All they had to do was follow Facebook's playbook and adopt some of the strategies they'd been so careful about avoiding, like sending more frequent notifications and suggestions to users about who else they should follow.

Those moves, some of which had seemed tasteless in the past, instantly sounded a lot more reasonable now that Instagram's trajectory was threatened. Instagram had long been able to scoff at Facebook's growth tactics, because Facebook had made growth easy for them. Ironically, in an act of competitive defiance against their own parent company, they ended up doing what Facebook had always advised.

· · · · · · · · · · · · · ·

In the chaos to try and reverse the slowdown, make IGTV work, and tussle with Facebook over resources, one group lost out the most: the team trying to make progress on solving Instagram's biggest problems before they became bigger scandals like Facebook's.

Every engineer wanted to prioritize building something new, in a company with data-based goal setting and growth as a priority. Those who worked on shutting down opioid sales or removing posts that glorified suicide had trouble measuring and being rewarded for their progress. After a hashtag was banned or a certain type of post taken down, users might organize their content under a new descriptor, or start talking about it in comments instead. How could you definitively measure the absence of something damaging, without knowing every time it appeared in billions of posts a day?

The well-being team, led by Ameet Ranadive, had been trying to teach a machine learning algorithm to identify comments that constituted bullying, so the comments could be automatically removed. But Ranadive wanted to move beyond bullying, and on to tackling 13 Instagram-specific issues, including drug sales and election interference.

Ranadive didn't know about Systrom's earlier conversations with the big Facebook integrity team. He only knew that Mosseri wasn't going to let him spend engineering resources on these problems. Mosseri was firm: in order to do his job successfully, he would have to think of areas to use Facebook's resources instead of hoping to increase Instagram's, wherever that would work.

"You need to stop what you're doing and drop everything to figure out a way to work with Facebook," he told Ranadive.

"Working with Facebook makes sense in theory, but we can't just stop," Ranadive said. The media was starting to write about Instagram's problems too. The *Washington Post* was planning a story about the sale of opioids though Instagram, and the communications team was asking Ranadive what his plan was. When it published in September, the *Post* explained that Instagram not only hosts drug content, but makes it easier for users to find sellers through its personalization.

"You just don't have the resources Facebook does to tackle this," Mosseri explained.

Ranadive appealed to Krieger, who tried to mediate the disagreement. Krieger, like Systrom, did help advocate for more attention to be paid to user well-being. But eventually, even Krieger admitted that Mosseri was right. Engineering resources were precious and Instagram was understaffed. And if Instagram could prioritize convincing Facebook's engineers to work on the app's problems, then Instagram's best people could work on new products that would help the app grow.

For Facebook, those problems would always seem like a side project.

And so at Instagram, which always said it prioritized community above all else, the community lost.

· · · · · · · · · · · · •

Systrom was expected to be back from parental leave at the end of July. And then he extended his leave to the end of August. And then the end of September. Through it all, he was meeting with his mentors and with Krieger. Both founders were increasingly frustrated, agonizing about the events of the past few months.

When Systrom returned, on a Monday in late September, he and Krieger organized a meeting with their top staff in the South Park conference room. As Mosseri and others arrived, there were hugs and smiles to celebrate Systrom's return at a stressful time.

And then Systrom told them he was resigning. So was Krieger.

At first the other leaders thought they were joking. They couldn't imagine the possibility of Instagram without the pair. But it was true. The founders had already told Cox, Zuckerberg, and Sandberg.

"We just think it's time," Systrom said. "We've thought about this a lot. We've been talking a lot." It had been six years, longer than anyone expected they'd last. They said they wanted to take a break and get back to their creative roots.

With their executive team, Systrom and Krieger were diplomatic about their reasons because they didn't want to stir drama. But they'd been quite clear with Cox that morning.

"Remember that conversation from earlier this year?" Systrom said to Cox. He'd asked for resources, independence, and trust. "None of the things I asked for have happened."

There was no plan for this scenario. Instagram and Facebook didn't have an internal communications strategy, an external communications strategy, a plan for succession, or a timeline for interviewing candidates. That was what Mosseri thought about as he realized that soon, everyone would find out, and everyone would start fretting externally as much as he was fretting in his head.

But there wasn't much time. His day was full of back-to-back meetings, in which he had to pretend nothing had happened. He interviewed a candidate for a product manager role, and then had a getting-to-know-you chat with the European team, and on and on, until he finally was on the company shuttle home to San Francisco, cranking through his inbox, and then walking in the door.

He took his shoes off and started talking to his wife, Monica.

"Kevin and Mike are leaving," he said.

"Really? What does that mean for you?" she said.

"I don't know," he said.

Just then his phone dinged with a news alert from the *New York Times*: "Instagram's Co-Founders to Step Down from Company." Within minutes, the story was everywhere.

• • • • • • • • • • • • •

Systrom and Krieger scrambled to write a short three-paragraph message to employees that night, which they also decided to post on the Instagram blog:

> Mike and I are grateful for the last eight years at Instagram and six years with the Facebook team. We've grown from 13 people to over a thousand with offices around the world, all while building products used and loved by a community of over one billion. We're now ready for our next chapter.
>
> We're planning on leaving Instagram to explore our curiosity and creativity again. Building new things requires that we step back, understand what inspires us and match that with what the world needs; that's what we plan to do.
>
> We remain excited for the future of Instagram and Facebook in the coming years as we transition from leaders to two users in a billion. We look forward to watching what these innovative and extraordinary companies do next.

Hidden in the bland statement were two symbolic gestures. There was no mention of Zuckerberg. And they referred to Instagram as a separate company, which it had not been for six years.

· · · · · · · · · · · ·

Mosseri did interview for the job. But because of the leaks to the media, he couldn't tell anyone when he received it, even when family members kept calling, wanting to know if the speculation about his promotion was true. Mosseri had to lie to his own mother, telling her he hadn't heard any news yet.

Before the company announced Mosseri's promotion, he went to Systrom's house up on a hill in San Francisco to pose on a couch with him and Krieger. The press was writing about mounting tensions between Facebook and Instagram, so the founders needed to endorse Mosseri, reassuring users through the image that the app they knew and loved wouldn't be ruined. The head of communications took the picture with Mosseri's camera—they couldn't risk an outside photographer because of all of the news reports. In the picture they are all smiling; in their address to employees that day, on the 30th floor of a new San Francisco office still under renovation for Instagram, they would be facing a room full of red, teary eyes.

"Since we announced our departure, many people have asked us what we hope for the future of Instagram," Systrom said in his post announcing Mosseri was taking over. "To us, the most important thing is keeping our community—all of you—front and center in all that Instagram does."

Mosseri's title would be "head of Instagram." At Facebook Inc., there was room for only one CEO.

EPILOGUE

THE PRICE OF THE ACQUISITION

At the end of 2019, Instagram announced that it would stop letting people see the like count on other users' photos. Results from a months-long test of the change showed positive effects on behavior, though Instagram wouldn't say exactly what the effects were. The like hiding, Adam Mosseri explained, was intended to reduce the inadequacy users feel when they compare their success to others, "to try and make Instagram feel less pressurized, to make it less of a competition." The app also started telling users when they'd seen all the new posts in their feed, so they could stop scrolling. Both moves were praised by the media and celebrities. Instagram seemed to be standing up for the well-being of its community.

But there was another change occurring sans press release that sent an entirely different message. Users were being asked, with a pop-up in their Instagram app, if they would like more analytics on their performance. These extra charts and graphs—to see what age range their account was reaching, how many people unfollowed their account that week, or which posts were most popular—had long been available to influencers and brands on Instagram. Now regular people were being invited to use Instagram's free data tools too.

It became a joke in some teen circles, at first, to officially tell Instagram they were a "DJ" or "model" or "actor" in exchange for these analytics and a tongue-in-cheek job label on their profiles. And then, as more and more people clicked to accept the tools, it just seemed normal. Of course everyone wanted more data on their performance. Wasn't the point of Instagram to create posts that other people wanted to follow?

The tech industry's obsession with measurement and trend analysis, honed by Facebook to give the people what they want on their news feeds, seemed at first to be incompatible with an app based on art and creativity. But over the years, Facebook infused its ethos into Instagram. As Instagram became part of our culture, Facebook's culture of measurement did too. The line between person and brand is blurring. The hustle for growth and relevance, backed by data, is now the drumbeat of modern life online. No matter what Instagram does with like counts, our communication has become more strategic. Instagram has made us not only more expressive but also more self-conscious and performative.

The data helps us distill complex human emotions and relationships into something easier to process. We can roughly assume that followers are equivalent to a level of interest in our lives or brands. Likes equal good content. Comments equal someone caring about that content. But to turn these numbers into goals is to make the same mistake, individually, as Facebook made in their organization, when Mark Zuckerberg decided the top objective was to grow the social network's number of users and increase the amount of time they spent on the app. The growth mission gave employees purpose, but also blind spots, and an incentive to take shortcuts.

Just like Instagram's users will have trouble letting go of likes, Facebook will have trouble changing the motivations of its workforce. Zuckerberg says he now wants to measure the social network's progress in terms of meaningful conversations, and time *well spent*. The problem is, the growth still needs to come from somewhere. It is, after all, a corporation.

In the months after the Instagram founders left, their app was rebranded as "Instagram from Facebook." The group in charge of Instagram's direct messaging was transferred to report to the Facebook Messenger team. In late 2019, Zuckerberg made a cameo appearance at an Instagram-branded conference and took a selfie with the crowd. Internally at Facebook, he was talking about using Instagram to take on

TikTok, the Chinese app that had replaced Snapchat as the top threat to Facebook's dominance. The frequency of advertising on Instagram had increased. There were more notifications too, and more personalized recommendations about who to follow. Being part of the Facebook "family" meant making compromises to bolster the bottom line—and to account for the growth rate slowing down on the main social network.

That October, Instagram employees gathered around a cake.

"Happy birthday to you, happy birthday to you, happy birthday to Instagram . . ." a crowd of dozens sang at an office party in San Francisco. It had been nine years since the founders clicked to launch the app to the world. The cake was the kind that existed because of Instagram, with five different-colored layers and surprise rainbow sprinkles that spilled out of the center once it was sliced.

Systrom and Krieger weren't at the party. Systrom hadn't even been posting on his Instagram account. Actually, he'd been un-posting. The picture taken on the couch in his home with Krieger and Adam Mosseri, to signal a friendly transfer of power, no longer appeared on his feed. Both founders were trying to take their time to look inward and think about who they were without their jobs. Systrom learned to fly his own planes. Krieger became a father.

The executives standing by the rainbow cake, including Mosseri, had all formerly been working on the Facebook side of the company and understood that achieving harmony with the company meant putting ego aside and slowly ceding control. Despite all the changes, Mosseri was determined to prove to employees that he would maintain the same push-and-pull with Facebook that had helped Systrom and Krieger come to better conclusions about what to build, not just by doing what seemed obvious to Zuckerberg. Mosseri had been running public question-and-answer sessions on his Instagram stories every Friday, trying to improve the public's understanding of how Instagram works. The week of the birthday party, he posted again.

"The most important question we face is, are we good for people?" Mosseri wrote.

That question is looming larger than ever in public discourse. In the

U.K., Instagram has had to answer for the suicide of 14-year-old Molly Russell. Her father blamed the app after finding material related to self-harm and depression when looking at her account after her death. In the U.S., Facebook has had to answer questions in Congress about the drugs for sale on Instagram. After a Facebook executive testified that Instagram had worked harder to take down images and hashtags, the activist Eileen Carey confronted her privately about the drug transactions still spurred through photo comments.

Around the world, Instagram's biggest fans—the people who have become famous and rich through the app—have spoken out about how difficult it is to keep up appearances. Instagram has been privately advising its stars to stop trying so hard to be perfect, and start posting more raw and vulnerable content. They explain that perfection is no longer novel. Vulnerability now gets better engagement, because it's more relatable.

There are also the regulatory questions. Governments are waking up to the idea that the top alternative to Facebook is an app also owned by Facebook. The U.S. Federal Trade Commission and the Department of Justice are both probing whether Facebook is a monopoly, and revisiting the Instagram acquisition as part of their investigation.

A debate about whether Facebook has too much power—so much that the government should force Instagram to be a separate company—is a hot topic on the 2020 U.S. presidential campaign trail. Politicians and academics alike argue that Facebook inflicted damage on society by not keeping track of the ways its users worked to influence elections, recruit terrorists, live-stream mass shootings, spread medical misinformation, and scam people. Zuckerberg says that Facebook now spends more on "integrity" issues than Twitter makes in annual revenue. He spent the year reframing his company's biggest problems as problems with "technology" or with "social media" at large.

Mosseri's answer to the important question was perfect by Facebook standards: "Technology isn't good or bad—it just is," he wrote. "Social media is a great amplifier. We need to do all we can responsibly to magnify the good and address the bad."

But nothing "just is," especially Instagram. Instagram isn't designed to be a neutral technology, like electricity or computer code. It's an intentionally crafted experience, with an impact on its users that is not inevitable, but is the product of a series of choices by its makers about how to shape behavior. Instagram trained its users on likes and follows, but that wasn't enough to create the emotional attachment users have to the product today. They also thought about their users as individuals, through the careful curation of an editorial strategy, and partnerships with top accounts. Instagram's team is expert at amplifying "the good."

When it comes to addressing "the bad," though, employees are concerned the app is thinking in terms of numbers, not people. Facebook's top argument against a breakup is that its "family of apps" evolution will be better for users' safety. "If you want to prevent interference in elections, if you want to reduce the spread of hate speech on the platforms, we benefit massively from working together closely," Mosseri said. But in practice, Instagram-specific problems get attention only after Facebook's headline issues are addressed first. Employees explain that at Facebook, it simply seems logical. Every decision is about impacting the most people possible, and Facebook has more users than Instagram does.

It makes sense that a network of people would have human problems. But even problems that hit hundreds of thousands of people can appear statistically insignificant to such a large company. In many cases, Instagram doesn't know the full scale of its problems because it hasn't invested in proactive detection. Instagram will take down a rash of photos of illegal activity, or a network of folks buying and selling verification, and then the problems will resurface in a different way. They'll ban young people from seeing plastic surgery filters, but then not have a proper age verification system. It's like a beautifully decorated apartment complex, full of pests and leaks. It requires a patch-up here, a trap there, and an occasional deep clean for it to be hospitable to tenants. But the managers of the building don't have the resources to think about where the leaks are starting, or whether there's a structural problem, because their contractors have to remodel the much bigger Facebook building first.

In 2019, Instagram delivered about $20 billion in revenue, more than a quarter of Facebook's overall sales. Facebook's 2012 cash-and-stock offer was a historic bargain in corporate acquisition history. After the cannibalization study, Instagram is growing more in Facebook's image than ever. The study was supposed to make the choice about what to do with Instagram rational and logical; Instagram employees are concerned it was used to rationalize Zuckerberg taking more control over the product.

Systrom and Krieger sold Instagram to Facebook because they wanted it to be bigger, more relevant, with more longevity. "You should be able to take a chance and build something of value for the world that should be able to grow and be worth a lot, and use that to give back socially," Systrom explained to *New York* magazine. "We tried really hard to do that, to be a force for good." But 1 billion users later, the app they developed to have tremendous cultural influence has been mixed up in a corporate struggle over personality, pride, and priorities. If Facebook's history is any guide, the real cost of the acquisition will fall on Instagram's users.

ACKNOWLEDGMENTS

This book was created from the thoughts and memories of so many people. I'm grateful for every single meal, coffee meeting, phone call, and conference room sit-down. There are sources who marked their calendars with a half-hour slot, and then sat with me for two hours or three, or allowed me to walk with them as I scribbled in my notebook through the streets of San Francisco, or endured my tedious follow-up questions. To help a journalist is to take a leap of faith. Thank you to everybody who does.

A heartfelt thanks also goes to Stephanie Frerich, my editor, who championed this book so strongly that she took the project with her when she switched jobs to Simon & Schuster. She chose to make this a part of her career, and devoted herself to inspiring me. Pilar Queen, my agent, has been an incredible advocate not just for this project but for me, as a first-time author, helping me understand how to succeed.

I wouldn't have been introduced to Pilar or Stephanie—honestly, I wouldn't even know that I had the capacity to write a book—if it weren't for Brad Stone. Brad, an author and senior executive editor of our technology team at Bloomberg News, knew I was going to write a book about Instagram long before I did. He proposed this idea to me in December 2017, while I was still working on a *Bloomberg Businessweek* cover story about the app, which eventually formed the basis for my proposal. Throughout the project, even as Brad ran our global team and worked on his second Amazon book, he was always available for a chat whenever I needed advice. I wouldn't be the journalist I am today without his guidance and tireless support.

Thanks to publisher Jonathan Karp and everyone on the Simon & Schuster team for making this project a reality. Emily Simonson has been a devoted assistant editor, advising me through each step of the process. Pete Garceau is responsible for the stunning cover, Jackie Seow for the art direction, and Lewelin Polanco for the interior design. Marie Florio is the reason this book is available in so many languages all over the world. If you've seen buzz around this book, it's because of Larry Hughes in publicity and Stephen Bedford in marketing. Thanks also to Sherry Wasserman and Alicia Brancato in production, as well as managing editor Kimberly Goldstein and assistant managing editor Annie Craig. I also am grateful for Felice Javit at S&S and Jamie Wolf at Pelosi, Wolf, Effron & Spates for their advice.

Bloomberg News leadership has demonstrated incredible support for the project, especially considering that it meant their Facebook reporter would be distracted or away from the desk during a period of congressional hearings, federal investigations, and privacy scandals. Thank you to all my colleagues who stepped in at various points to write news about Facebook, including Selina Wang and Gerrit DeVynck. In the spring of 2019, Kurt Wagner joined Bloomberg as a second Facebook reporter, and had to get up to speed on the job while I was away writing this book. He did great work then, allowing me to focus, which felt like a wonderful gift. I'm lucky to work so closely with someone I trust.

I'm eternally grateful to Tom Giles and Jillian Ward for the way they work with Brad to lead our Bloomberg tech team, always advocating for their reporters' ideas and careers. The team is full of best-in-class journalists and editors from whom I get to learn every day. Jillian, Emily Biuso, and Alistair Barr read some draft chapters and gave feedback when I felt stuck on revisions while Anne VanderMey, Ellen Huet, Dina Bass, Shira Ovide, Mark Bergen, Austin Carr, and Kurt served as beta readers at the very end. Nico Grant, who sits next to me at the office, was a trusted confidant and friend throughout the process. Emily Chang and Ashlee Vance, both colleagues who have written amazing books, served as role models and provided counsel and support throughout the process. Max Chafkin has been the main editor of all my long-form writing for *Bloomberg Businessweek*, including a cover story about Instagram. Working with him over the years prepared me to write this.

Genevieve Grdina and Elisabeth Diana, on the Instagram communications team, were important advocates for this project within Facebook. Thank you to everyone at Facebook and Instagram who took time out of their busy schedules to sit down with me for this book, or answer my fact-checking questions. This book is more accurate because of their participation. Thank you to the influencers and small businesses who shared their stories, especially those in São Paulo, who let me see what their work is like behind the scenes. I learned so much from all of the people who have turned Instagram into their career. Thank you especially to Instagram's founders, without whom none of this would be possible. You created something that truly changed the world.

I'm grateful for Sean Lavery, who fact-checked this manuscript and also reassured me during a stressful time. Jessica J. Lee drafted the endnotes, and Blake Montgomery compiled research on Instagram's cultural impact for me when I was in the brainstorming stages. Shruti Shah, Alexia Bonastos, and Sarah Seegal were friendly faces at The Wing when I was in my most desperate writing stages.

I am a business journalist in the first place because of Chris Roush and Penelope Abernathy, who at UNC taught me to think critically about how corporations work. Penny told me I'd be writing a book within five years of graduation. Sorry to miss the deadline!

If journalism is the first rough draft of history, books build on that important work to propose a second draft. I'm grateful to all the reporters who asked questions of Instagram throughout the years, and to the people who continue to cover its impact on our society and culture, and its place within Facebook. Reporters have been cited in the endnotes if their work found a second life in these pages.

The journalism community has been supportive in other ways too. Other authors, including Nick Bilton, Blake Harris, and Roger McNamee, reached out and offered help in key moments. Tim Higgins and

Alex Davies, friends who were writing their books at the same time, were important sounding boards and therapeutic dinner companions. Kara Swisher, a mentor to so many young journalists in Silicon Valley, advocated for this project and introduced me to fascinating people who made these chapters richer.

This process has made me so grateful to be surrounded by such brilliant, kind, selfless friends and family. Claire Korzen, my wonderful cousin who has written stories and silly plays with me since childhood, was the first person to read this book and provided invaluable feedback and encouragement on the rawest draft, when I was feeling the most vulnerable. My friend Keicy Tolbert read after Claire, and provided incredibly thoughtful high-level commentary that shaped my revisions. My cousin Michelle Kolodin introduced me to friends who were heavy Instagram users, all of whom had interesting perspectives. Walter Hickey marked up one of my chapters on a plane flight, while Owen Thomas made sure I had my tech history right. Ashley Lutz and Katie Ho provided welcome beachside distraction, while Will Bondurant ensured I didn't miss any UNC basketball. Miranda Henely sent me a thoughtful package to celebrate the project and encourage relaxation. Alex Barinka was constantly brainstorming on my behalf. Christina Farr lovingly forced me to read sections of this book out loud to her on her couch over glasses of wine. She then interrogated me about the things that didn't make sense, as all good journalists do, and generally made sure I was okay.

My younger brother, Michael Frier, kindly sent me pages of research about Instagram's impact on mental health, and was so enthusiastic about this project. My older brother, James Frier, and his wife, Maddie Tuller Frier, were also wonderful, and hosted me on their couch during reporting trips to Los Angeles. Over Christmas, the whole family helped me, especially Maddie, who caught dozens of typos.

I would not have finished this book as quickly or thoughtfully without my dad, Ken Frier, and his wife, Gretchen Tai, who opened up their home to me. They cooked me delicious meals and generally let me focus when I was at the very end of my deadline and unable to work in my apartment. My dad also lent a sharp eye to some sentences when I was stuck. For that I must thank his parents, John and Mary Ellen Frier, who inspired a couple generations of voracious readers and problem solvers.

My mom, Laura Casas, besides being so generally supportive and encouraging, helped my husband and me move apartments in the middle of all this. She also was able to help me communicate with my abuelita, Gudelia Casas, who was sick. My abuelita, who knew no English when she immigrated to this country in 1956 with small children, lived long enough to see the first printed version of this book. Her courage and kindness continues to inspire me.

And most of all, thank you to my love, Matt, for being there for me every day, providing strength, inspiration, and even the occasional delicious pastry. You make everything possible, and this book is dedicated to you.

NOTES

1 | PROJECT CODENAME

1   *"I like to say I'm dangerous . . . entrepreneurship."*: Charlie Parrish, "Instagram's Kevin Systrom: 'I'm Dangerous Enough to Code and Sociable Enough to Sell Our Company,'" *The Telegraph*, May 1, 2015, https://www.telegraph.co.uk/technology/11568119/Instagrams-Kevin-Systrom-Im-dangerous-enough-to-code-and-sociable-enough-to-sell-our-company.html.

4   *too rule-abiding to drink there*: Kevin Systrom, "How to Keep It Simple While Scaling Big," interview by Reid Hoffman, *Masters of Scale*, podcast audio, accessed September 7, 2018, https://mastersofscale.com/kevin-systrom-how-to-keep-it-simple-while-scaling-big/.

4   *also been the captain of the lacrosse team*: Parrish, "Instagram's Kevin Systrom."

4   *"to show my outlook . . . new way."*: D. C. Denison, "Instagram Cofounders' Success Story Has Holliston Roots," *Boston Globe*, April 11, 2012, https://www.bostonglobe.com/business/2012/04/11/instagram-cofounder-success-story-has-holliston-roots/PzCxOXWFtfoyWYfLKRM9bL/story.html.

4   *"You're not here to do perfection"*: Systrom, "How to Keep It Simple."

5   *Systrom needed a startup internship*: Kevin Systrom, "Tactics, Books, and the Path to a Billion Users," interview by Tim Ferriss, *The Tim Ferriss Show*, podcast audio, accessed September 7, 2018, https://tim.blog/2019/04/30/the-tim-ferriss-show-transcripts-kevin-systrom-369/.

5   *recapturing the abandoned real estate*: Michael V. Copeland and Om Malik, "Tech's Big Comeback," *Business 2.0 Magazine*, January 27, 2006, https://archive.fortune.com/magazines/business2/business2_archive/2005/11/01/8362807/index.htm.

6   *He would sometimes dream*: Nick Bilton, *Hatching Twitter: A True Story of Money, Power, Friendship, and Betrayal* (New York: Portfolio, 2014), 121.

6   *There were only a handful of employees*: Murad Ahmed, "Meet Kevin Systrom: The Brain Behind Instagram," *The Times*, October 5, 2013, https://www.thetimes.co.uk/article/meet-kevin-systrom-the-brain-behind-instagram-p5kvqmnhkcl.

7   *His last year at Stanford*: Steven Bertoni, "Instagram's Kevin Systrom: The Stanford Billionaire Machine Strikes Again," *Forbes*, August 1, 2012, https://www.forbes.com/sites/stevenbertoni/2012/08/01/instagrams-kevin-systrom-the-stanford-millionaire-machine-strikes-again/#36b4306d45b9.

8   They're crazy, *Systrom thought*: Kevin Systrom, "Billion Dollar Baby," interview by Sarah Lacy, Startups.com, July 24, 2017, https://www.startups.com/library/founder-stories/kevin-systrom.

8   *He'd have a salary with a base of about $60,000*: Bertoni, "Instagram's Kevin Systrom."

8   *A slightly purpler blue shade*: Alex Hern, "Why Google Has 200M Reasons to Put Engineers over Designers," *The Guardian*, February 5, 2014, https://www.theguardian.com/technology/2014/feb/05/why-google-engineers-designers.

10   *Big web services like Facebook*: Jared Newman, "Whatever Happened to the Hottest iPhone Apps of 2009?," *Fast Company*, May 31, 2019, https://www.fastcompany.com/90356079/whatever-happened-to-the-hottest-iphone-apps-of-2009.

11   *It would be swarming with venture capitalists*: Stewart Butterfield and Caterina Fake, "How We Did It: Stewart Butterfield and Caterina Fake, Cofounders, Flickr," *Inc.*, December 1, 2006, https://www.inc.com/magazine/20061201/hidi-butterfield-fake.html.

11   *Chris Dixon, who'd sold a security*: Chris Dixon, author biography, Andreesen Horowitz, accessed September 18, 2019, https://a16z.com/author/chris-dixon/.

14   *according to Nick Bilton's book*: Bilton, *Hatching Twitter*, 120.

15   *froze the Facebook bank account*: Nicholas Carlson, "Here's the Email Zuckerberg Sent to Cut His Cofounder Out of Facebook," *Business Insider*, May 15, 2012, https://www.businessinsider.com/exclusive-heres-the-email-zuckerberg-sent-to-cut-his-cofounder-out-of-facebook-2012-5?IR=T.

16   *Systrom thought of how scary*: Systrom, "Tactics, Books, and the Path to a Billion Users."

18   *"I think there will be . . . around these phones."*: "Full Transcript: Instagram CEO Kevin Systrom on Recode Decode," *Vox*, June 22, 2017, https://www.vox.com/2017/6/22/15849966/transcript-instagram-ceo-kevin-systrom-facebook-photo-video-recode-decode.

20   *"You know what he does to those photos, right?"*: Kara Swisher, "The Money Shot," *Vanity Fair*, May 6, 2013, https://www.vanityfair.com/news/business/2013/06/kara-swisher-instagram.

21 *Hipstamatic, with which you could make your photos*: M. G. Siegler, "Apple's Apps of the Year: Hipstamatic, Plants vs. Zombies, Flipboard, and Osmos," *TechCrunch*, December 9, 2010, https://techcrunch.com/2010/12/09/apple -top-apps-2010/.

26 *Everything he posted on Instagram*: Steve Dorsey (@dorsey), "@HartleyAJ Saw that and thought it was remarkable (but wasn't sure what to call it). Thanks, WX-man! :)," Twitter, November 9, 2010, https://web.archive.org /web/20101109211738/http://twitter.com/dorsey.

2 | THE CHAOS OF SUCCESS

30 *"Instagram was so . . . simple."*: Dan Rubin, interview with the author, phone, February 8, 2019.

35 *By January, brands like Pepsi*: M. G. Siegler, "Beyond the Filters: Brands Begin to Pour into Instagram," *TechCrunch*, January 13, 2011, https://techcrunch .com/2011/01/13/instagram-brands/?_ga=2.108294978.135876931.15598 87390-830531025.1555608191.

35 *"We're not interested . . . use the product."*: Siegler, "Beyond the Filters."

36 *In his estimation, Snoop*: M. G. Siegler, "Snoopin' on Instagram: The Early-Adopting Celeb Joins the Photo-Sharing Service," *TechCrunch*, January 19, 2011, https://techcrunch.com/2011/01/19/snoop-dogg-instagram/.

38 *"Justin Bieber Joins Instagram, World Explodes"*: Chris Gayomali, "Justin Bie-ber Joins Instagram, World Explodes," *Time*, July 22, 2011, http://techland .time.com/2011/07/22/justin-bieber-joins-instagram-world-explodes/.

41 *They called the process "pruning the trolls"*: Nicholas Thompson, "Mr. Nice Guy: Instagram's Kevin Systrom Wants to Clean Up the &#%$@! Internet," *Wired*, August 14, 2017, https://www.wired.com/2017/08/instagram-kevin-systrom -wants-to-clean-up-the-internet/.

41 *After just nine months, the app*: M. G. Siegler, "The Latest Crazy Instagram Stats: 150 Million Photos, 15 per Second, 80% Filtered," *TechCrunch*, Au-gust 3, 2011, https://techcrunch.com/2011/08/03/instagram-150-million/.

41 *According to Section 230 of the Communications Decency Act*: Protection for private blocking and screening of offensive material, 47 U.S. Code § 230 (1996).

3 | THE SURPRISE

50 *"He chose us, not the other way around."*: Dan Rose, interview with the author, Facebook headquarters, December 18, 2018.

53 *Google had bought YouTube for $1.6 billion*: Associated Press, "Google Buys YouTube for $1.65 Billion," *NBC News*, October 10, 2006, http://www.nbc

news.com/id/15196982/ns/business-us_business/t/google-buys-youtube -billion/#.XX9Q96d7Hox.

54   *One billion dollars, Reuters said*: Alexei Oreskovic and Gerry Shih, "Facebook to Buy Instagram for $1 Billion," Reuters, April 9, 2012, https://www.reuters.com/article/us-facebook/facebook-to-buy-instagram-for-1-billion-idUS-BRE8380M820120409.

54   *Zuckerberg was "paying a steep . . . model"*: Laurie Segall, "Facebook Acquires Instagram for $1 billion," *CNN Money*, April 9, 2012, https://money .cnn.com/2012/04/09/technology/facebook_acquires_instagram/index.htm.

57   *But it would take some serious negotiating*: Shayndi Raice, Spencer E. Ante, and Emily Glazer, "In Facebook Deal, Board Was All but Out of Picture," *Wall Street Journal*, April 18, 2012, https://www.wsj.com/articles/SB10001424 052702304818404577350191931921290.

57   *started out by asking for $2 billion:* Raice, Ante, and Glazer, "In Facebook Deal, Board Was All but Out of Picture."

60   *The discussions continued at Zuckerberg's*: Mike Swift and Pete Carey, "Facebook's Mark Zuckerberg Buys House in Palo Alto," *Mercury News*, May 4, 2011, https://www.mercurynews.com/2011/05/04/facebooks-mark-zuckerberg -buys-house-in-palo-alto/.

61   *But if it's a bubble*: Aileen Lee, "Welcome to the Unicorn Club, 2015: Learning from Billion-Dollar Companies," *TechCrunch*, July 18, 2015, https:// techcrunch.com/2015/07/18/welcome-to-the-unicorn-club-2015-learning -from-billion-dollar-companies/.

68   *"The 13 employees of . . . multi-millionaires"*: Julian Gavaghan and Lydia Warren, "Instagram's 13 Employees Share $100M as CEO Set to Make $400M Reveals He Once Turned Down a Job at Facebook," *Daily Mail*, April 9, 2012, https://www.dailymail.co.uk/news/article-2127343/Facebook-buys-Instagram -13-employees-share-100m-CEO-Kevin-Systrom-set-make-400m.html.

68   *"Instagram is now worth $77 million"*: Derek Thompson, "Instagram Is Now Worth $77 Million per Employee," *The Atlantic*, April 9, 2012, https://www .theatlantic.com/business/archive/2012/04/instagram-is-now-worth-77 -million-per-employee/255640/.

68   Business Insider *published a list*: Alyson Shontell, "Meet the 13 Lucky Employees and 9 Investors Behind $1 Billion Instagram," *Business Insider*, April 9, 2012, https://www.businessinsider.com/instagram-employees-and -investors-2012-4?IR=T.

4 | THE SUMMER IN LIMBO

69   *"I write to urge . . . social network market."*: David Cicilline, "Cicilline to FTC—Time to Investigate Facebook," March 19, 2019, https://cicilline

.house.gov/press-release/cicilline-ftc-%E2%80%93-time-investigate
-facebook.

74   *Shareholders brought a class-action lawsuit*: Jonathan Stempel, "Facebook
     Settles Lawsuit Over 2012 IPO for $35 Million," Reuters, February 26, 2018,
     https://www.reuters.com/article/us-facebook-settlement/facebook-settles
     -lawsuit-over-2012-ipo-for-35-million-idUSKCN1GA2JR.

74   *Around the world, members of the social network*: Danielle Kucera and Doug-
     las MacMillan, "Facebook Investor Spending Month's Salary Exposes Hype,"
     Bloomberg.com, May 24, 2012, https://www.bloomberg.com/news/articles
     /2012-05-24/facebook-investor-spending-month-s-salary-exposes-hype.

76   *Other apps, like Camera Awesome*: Josh Constine, "FB Launches Facebook
     Camera: An Instagram-Style Photo Filtering, Sharing, Viewing iOS App," *Tech-
     Crunch*, May 24, 2012, https://techcrunch.com/2012/05/24/facebook-camera/.

76   *The Office of Fair Trading wrote*: UK Office of Fair Trading, "Anticipated
     Acquisition by Facebook Inc of Instagram Inc," August 22, 2012, https://
     webarchive.nationalarchives.gov.uk/20140402232639/http://www.oft.gov.uk
     /shared_oft/mergers_ea02/2012/facebook.pdf.

76   *which had fewer than 3 million users:* Matthew Panzarino, "Dave Morin: Path to
     Hit 3M Users This Week, Will Release iPad App This Year, But Not For Windows
     Phone," *The Next Web*, June 1, 2012, https://thenextweb.com/apps/2012/06/01
     /dave-morin-path-to-hit-3m-users-this-week-will-release-ipad-app-this-year/.

77   *Path shut down in 2018*: Harrison Weber, "Path, the Doomed Social Network
     with One Great Idea, Is Finally Shutting Down," *Gizmodo*, September 17,
     2018, https://gizmodo.com/path-the-doomed-social-network-with-one-great
     -idea-is-1829106338.

77   *selling to a South Korean company, Daum Kakao*: Edwin Chan and Sarah Frier,
     "Morin Sells Chat App Path to South Korea's Daum Kakao," Bloomberg
     .com, May 29, 2015, https://www.bloomberg.com/news/articles/2015-05-29
     /path-s-david-morin-sells-chat-app-to-south-korea-s-daum-kakao.

78   *It was Facebook's job to not let anyone*: Evan Osnos, "Can Mark Zuckerberg Fix
     Facebook Before It Breaks Democracy?," *New Yorker*, September 10, 2018,
     https://www.newyorker.com/magazine/2018/09/17/can-mark-zuckerberg
     -fix-facebook-before-it-breaks-democracy.

78   *Analysts would later say that approving*: Kurt Wagner, "Facebook's Acquisi-
     tion of Instagram Was the Greatest Regulatory Failure of the Past Decade,
     Says Stratechery's Ben Thompson," *Vox*, June 2, 2018, https://www.vox.com
     /2018/6/2/17413786/ben-thompson-facebook-google-aggregator-platform
     -code-conference-2018.

78   *"Mark's power is unprecedented and un-American"*: Chris Hughes, "It's Time
     to Break Up Facebook," *New York Times*, May 9, 2019, https://www.nytimes
     .com/2019/05/09/opinion/sunday/chris-hughes-facebook-zuckerberg.html#.

78   *The letters included a caveat*: April J. Tabor (US Federal Trade Commission), "Letter to Thomas O. Barnett," August 22, 2012, https://www.ftc.gov/sites/default/files/documents/closing_letters/facebook-inc./instagram-inc./120822barnettfacebookcltr.pdf.

79   *Every company that had built its servers*: Robert McMillan, "(Real) Storm Crushes Amazon Cloud, Knocks Out Netflix, Pinterest, Instagram," *Wired*, June 30, 2012, https://www.wired.com/2012/06/real-clouds-crush-amazon/.

81   *"If you go on the 'Popular' . . . drive it."*: Jamie Oliver and Kevin Systrom, "Jamie Oliver & Kevin Systrom, with Loic Le Meur - LeWeb London 2012 - Plenary 1," June 20, 2012, YouTube video, 32:33, https://www.youtube.com/watch?v=Pdbzmk0xBW8.

82   *"While we're excited . . . new users."*: Kris Holt, "Instagram Shakes Up Its Suggested Users List," *Daily Dot*, August 13, 2012, https://www.dailydot.com/news/instagram-suggested-users-shakeup/.

83   *"The companies and brands . . . genuine."*: Oliver and Systrom, "Jamie Oliver & Kevin Systrom, with Loic Le Meur."

84   *"We understand . . . no longer available within Instagram"*: Brian Anthony Hernandez, "Twitter Confirms Removing Follow Graph from Instagram's 'Find Friends'," *Mashable*, July 27, 2012, https://mashable.com/2012/07/27/twitter-instagram-find-friends/?europe=true.

5 | MOVE FAST AND BREAK THINGS

87   *"I hate when people discount us . . . wrong."*: Systrom, "Tactics, Books, and the Path to a Billion Users."

99   *"Instagram says . . . sell your photos"*: Declan McCullagh, "Instagram Says It Now Has the Right to Sell Your Photos," *CNET*, December 17, 2012, https://www.cnet.com/news/instagram-says-it-now-has-the-right-to-sell-your-photos/.

99   *"Facebook forces . . . their uploaded photos"*: Charles Arthur, "Facebook Forces Instagram Users to Allow It to Sell Their Uploaded Photos," *The Guardian*, December 18, 2012, https://www.theguardian.com/technology/2012/dec/18/facebook-instagram-sell-uploaded-photos.

100 *"Instagram users . . . photos are your photos"*: Instagram, "Thank You, and We're Listening," December 18, 2012, Tumblr post, https://instagram.tumblr.com/post/38252135408/thank-you-and-were-listening.

6 | DOMINATION

108 *"I have a special machine for it . . . graph."*: Dan Rookwood, "The Many Stories of Instagram's Billionaire Founder," *MR PORTER*, accessed May 2019,

https://www.mrporter.com/en-us/journal/the-interview/the-many-stories-of
-instagrams-billionaire-founder/2695.

108 *He once lost to a friend's teenage daughter*: Osnos, "Can Mark Zuckerberg Fix Facebook?"

108 "Carthago delenda est!": Antonio García Martínez, "How Mark Zuckerberg Led Facebook's War to Crush Google Plus," *Vanity Fair*, June 3, 2016, https:// www.vanityfair.com/news/2016/06/how-mark-zuckerberg-led-facebooks -war-to-crush-google-plus.

110 *It was "an artistic choice,"*: Colleen Taylor, "Instagram Launches 15-Second Video Sharing Feature, with 13 Filters and Editing," *TechCrunch*, June 20, 2013, https://techcrunch.com/2013/06/20/facebook-instagram-video/.

113 *"It's the fastest way . . . photos that disappear"*: Rob Price and Alyson Shontell, "This Fratty Email Reveals How CEO Evan Spiegel First Pitched Snapchat as an App for 'Certified Bros'," *Insider*, February 3, 2017, https://www.insider.com /snap-ceo-evan-spiegel-pitched-snapchat-fratty-email-2011-certified-bro-2017-2.

113 *He was the son of a powerful corporate*: John W. Spiegel, professional biography, Munger, Tolles & Olson, accessed February 12, 2018, https://www.mto.com /lawyers/john-w-spiegel.

113 *Besides having a tendency for profanity*: Sam Biddle, "'Fuck Bitches Get Leid': The Sleazy Frat Emails of Snapchat's CEO," *Valleywag*, May 28, 2014, http://valleywag.gawker.com/fuck-bitches-get-leid-the-sleazy-frat-emails-of -snap-1582604137.

113 *"People are living with this massive . . . communicating."*: J. J. Colao, "Snapchat: The Biggest No-Revenue Mobile App Since Instagram," *Forbes*, November 27, 2012, https://www.forbes.com/sites/jjcolao/2012/11/27/snapchat-the-biggest -no-revenue-mobile-app-since-instagram/#6ef95f0a7200.

114 *By November 2012, Snapchat*: Colao, "Snapchat."

115 *"Thanks :) would be happy . . . Bay Area,"*: Alyson Shontell, "How Snapchat's CEO Got Mark Zuckerberg to Fly to LA for Private Meeting," *Business Insider*, January 6, 2014, https://www.businessinsider.com/evan-spiegel-and -mark-zuckerbergs-emails-2014-1?IR=T.

115 *He spent the meeting insinuating*: J. J. Colao, "The Inside Story of Snapchat: The World's Hottest App or a $3 Billion Disappearing Act?," *Forbes*, January 20, 2014, https://www.forbes.com/sites/jjcolao/2014/01/06/the-inside-story -of-snapchat-the-worlds-hottest-app-or-a-3-billion-disappearing-act/.

116 *And then, starting the next day*: Seth Fiegerman, "Facebook Poke Falls Out of Top 25 Apps as Snapchat Hits Top 5," *Mashable*, December 26, 2012, https:// mashable.com/2012/12/26/facebook-poke-app-ranking/.

116 *Snapchat's downloads climbed*: Fiegerman, "Facebook Poke Falls Out of Top 25 Apps."

116   *In June 2013, Spiegel raised*: Mike Isaac, "Snapchat Closes $60 Million Round Led by IVP, Now at 200 Million Daily Snaps," *All Things D*, June 24, 2013, http://allthingsd.com/20130624/snapchat-closes-60-million-round-led-by -ivp-now-at-200-million-daily-snaps/.

118   *That September, Emily White was*: Evelyn M. Rusli, "Instagram Pictures Itself Making Money," *Wall Street Journal*, September 8, 2013, https://www.wsj .com/articles/instagram-pictures-itself-making-money-1378675706.

119   *Instagram ran its very first*: Kurt Wagner, "Instagram's First Ad Hits Feeds Amid Mixed Reviews," *Mashable*, November 1, 2013, https://mashable .com/2013/11/01/instagram-ads-first/.

119   *"5:15 PM: Pampered in Paris #MKTimeless"*: Michael Kors (@michaelkors), "5:15 PM: Pampered in Paris #MKTimeless," Instagram, November 1, 2013, https://www.instagram.com/p/gLYVDzHLvn/?hl=en.

122   *The company had Instagram reduce*: Dom Hofmann (@dhof), "ig blocked the #vine hashtag during our first few months," Twitter, September 23, 2019, 4:14 p.m., https://twitter.com/dhof/status/1176137843720314880.

122   *discouraged prominent users . . . Snapchat usernames*: Georgia Wells and Deepa Seetharaman, "Snap Detailed Facebook's Aggressive Tactics in 'Project Voldemort' Dossier," *Wall Street Journal*, last modified September 24, 2019, https://www.wsj.com/articles/snap-detailed-facebooks-aggressive-tactics -in-project-voldemort-dossier-11569236404.

124   *About 40 ideas emerged*: Brad Stone and Sarah Frier, "Facebook Turns 10: The Mark Zuckerberg Interview," Bloomberg.com, January 31, 2014, https:// www.bloomberg.com/news/articles/2014-01-30/facebook-turns-10-the-mark -zuckerberg-interview#p2.

7 | THE NEW CELEBRITY

126   *"There are plenty . . . phenomenon."*: Guy Oseary, inverview with the author, phone, March 20, 2019.

127   *Her dining room was decorated*: Madeline Stone, "Randi Zuckerberg Has Sold Her Boldly Decorated Los Altos Home for $6.55 Million," *Business Insider*, June 15, 2015, https://www.businessinsider.com/randi-zuckerberg -sells-house-for-655-million-2015-6?IR=T.

128   *She left the company before its initial*: Kara Swisher, "Exclusive: Randi Zuckerberg Leaves Facebook to Start New Social Media Firm (Resignation Letter)," *All Things D*, August 3, 2011, http://allthingsd.com/20110803/exclusive -randi-zuckerberg-leaves-facebook-to-start-new-social-media-firm-resignation -letter/.

130   *"As I brought him . . . fazed by anything,"*: Erin Foster, interview with the author, phone, July 16, 2019.

135 *"A lot of people thought . . . interesting,"*: Kris Jenner, interview with the author, phone, May 21, 2019.

136 *On the reality show* The Simple Life: *Access Hollywood*, "Paris Hilton on the Public's Misconception of Her & More (Exclusive)," YouTube video, 3:07, November 30, 2016, https://www.youtube.com/watch?v=ZqqAkp8zKp8&feature=youtu.be.

136 *"I started thinking, If Barbie . . . like?"*: Jason Moore, interview with the author, phone, April 21, 2019.

137 *"Then the publication . . . were behind it,"*: Moore, interview with the author.

137 *"We were used to getting paid . . . photo,"*: Moore, interview with the author.

138 *Jenner realized, "It's fun . . . join the party."*: Jenner, interview with the author.

138 *Since consumers are much more likely*: "Recommendations from Friends Remain Most Credible Form of Advertising Among Consumers; Branded Websites Are the Second-Highest-Rated Form," Nielsen N.V., September 28, 2015, https://www.nielsen.com/eu/en/press-releases/2015/recommendations-from-friends-remain-most-credible-form-of-advertising/.

145 *"Who's to say the person . . . stage."*: Darren Heitner, "Instagram Marketing Helped Make This Multi-Million Dollar Nutritional Supplement Company," *Forbes*, March 19, 2014, https://www.forbes.com/sites/darrenheitner/2014/03/19/instagram-marketing-helped-make-this-multi-million-dollar-nutritional-supplement-company/#4b317f2f1f2c.

146 *"We were used to going . . . magazines,"*: Christopher Bailey, interview with the author, phone, May 15, 2019.

147 *He argued that the brand couldn't*: Bailey, interview with the author.

152 *Twitter's television partnerships group*: Fred Graver, "The True Story of the 'Ellen Selfie,'" *Medium*, February 23, 2017, https://medium.com/@fredgraver/the-true-story-of-the-ellen-selfie-eb8035c9b34d.

152 *During rehearsal, DeGeneres saw*: Graver, "The True Story of the 'Ellen Selfie.'"

152 *The team presented her with*: Ibid.

156 *It featured Joan Smalls, Cara Delevingne*: "The Instagirls: Joan Smalls, Cara Delevingne, Karlie Kloss, and More on the September Cover of *Vogue*," *Vogue*, August 18, 2014, https://www.vogue.com/article/supermodel-cover-september-2014.

156 *"The girls were using Instagram . . . connected."*: Anna Wintour, interview with the author, phone, March 20, 2019.

157 *Twitter executives would say they*: Josh Halliday, "Twitter's Tony Wang: 'We Are the Free Speech Wing of the Free Speech Party,'" *The Guardian*, March 22, 2012, https://www.theguardian.com/media/2012/mar/22/twitter-tony-wang-free-speech.

157 *"If you think about the impact Twitter . . . pictures."*: Erin Griffith, "Twitter Co-Founder Evan Williams: 'I Don't Give a Shit' if Instagram Has More

Users," *Fortune*, December 11, 2014, https://fortune.com/2014/12/11/twitter
-evan-williams-instagram/.

158 *More than 50 stars*: "See Mark Seliger's Instagram Portraits from the 2015
Oscar Party," *Vanity Fair*, February 23, 2015, https://www.vanityfair.com
/hollywood/2015/02/mark-seliger-oscar-party-portraits-2015.

8 | THE PURSUIT OF THE INSTA-WORTHY

161 *After Jenner shared DuShane's Instagram*: Casey Lewis, "Kylie Jenner Just
Launched an Anti-Bullying Campaign, and We Talked to Her First Star,"
Teen Vogue, September 1, 2015, https://www.teenvogue.com/story/kylie
-jenner-anti-bullying-instagram-campaign.

165 *In February 2015, Twitter paid*: Peter Kafka, "Twitter Buys Niche, a Social
Media Talent Agency, for at Least $30 Million," *Vox*, February 11, 2015,
https://www.vox.com/2015/2/11/11558936/twitter-buys-niche-a-social-media
-talent-agency.

168 *"I was leading locals from . . . city,"*: Edward Barnieh, interview with the author, phone, June 7, 2019.

168 *"They knew Instagram . . . way."*: Barnieh, interview with the author.

169 National Geographic *wrote about how Instagram*: Carrie Miller, "How
Instagram Is Changing Travel," *National Geographic*, January 26, 2017,
https://www.nationalgeographic.com/travel/travel-interests/arts-and-culture
/how-instagram-is-changing-travel/.

169 *The caption is a simple hashtag*: Lucian Yock Lam (@yock7), "#Followme
bro," Instagram, December 16, 2015, https://www.instagram.com/p/_WhC
G7ISWd/?hl=en.

174 *The media dubbed the event*: Taylor Lorenz, " 'Instagram Rapture' Claims
Millions of Celebrity Instagram Followers," *Business Insider*, December 18,
2014, https://www.businessinsider.com/instagram-rapture-claims-millions-of
-celebrity-instagram-followers-2014-12.

175 Bloomberg Businessweek *reporter Max Chafkin tested*: Max Chafkin, "Con-
fessions of an Instagram Influencer," *Bloomberg Businessweek*, November
30, 2016, https://www.bloomberg.com/news/features/2016-11-30/confessions
-of-an-instagram-influencer.

9 | THE SNAPCHAT PROBLEM

179 *"What people are experiencing . . . popularity."*: Sean Burch, "Snapchat's Evan
Spiegel Says Instagram 'Feels Terrible' to Users," *The Wrap*, November 1,
2018, https://www.thewrap.com/evan-spiegel-snap-instagram-terrible/.

188 *In the fall of 2015, Ira Glass*: Ira Glass, "Status Update," *This American Life*,

November 27, 2015, https://www.thisamericanlife.org/573/status-update. Used with permission.

189 *As they explained to Glass*: Glass, "Status Update."

192 *"While we adored . . . Snapchat."*: Kendall Fisher, "What You Didn't See at the 2016 Oscars: Kate Hudson, Nick Jonas, Lady Gaga and More Take Us Behind the Scenes on Snapchat," *E! News*, February 29, 2016, https://www.eonline.com /fr/news/744642/what-you-didn-t-see-at-the-2016-oscars-kate-hudson-nick -jonas-lady-gaga-and-more-take-us-behind-the-scenes-on-snapchat.

196 *"Pray for me,"*: Pope Francis (@franciscus), "Pray for me," Instagram, March 19, 2016, https://www.instagram.com/p/BDIgGXqAQsq/?hl=en.

197 *"What this is about . . . possible."*: Mike Isaac, "Instagram May Change Your Feed, Personalizing It with an Algorithm," *New York Times*, March 15, 2016, https://www.nytimes.com/2016/03/16/technology/instagram-feed.html.

200 *His employees saw him as stubborn*: Brad Stone and Sarah Frier, "Evan Spiegel Reveals Plan to Turn Snapchat into a Real Business," *Bloomberg Businessweek*, May 26, 2015, https://www.bloomberg.com/news/features /2015-05-26/evan-spiegel-reveals-plan-to-turn-snapchat-into-a-real-business.

206 *His caption told the world his time*: Kevin Systrom (@kevin), "The last cycling climb of our vacation was the infamous Mont Ventoux," Instagram, August 17, 2016, https://www.instagram.com/p/BJN3MKIhAjz/?hl=en.

10 | CANNIBALIZATION

209 *Most Instagram users had no idea*: Casey Newton, "America Doesn't Trust Facebook," *The Verge*, October 27, 2017, https://www.theverge .com/2017/10/27/16552620/facebook-trust-survey-usage-popularity-fake-news.

211 *the top stories with false information*: Craig Silverman, "This Analysis Shows How Viral Fake Election News Stories Outperformed Real News on Facebook," *BuzzFeed News*, November 16, 2016, https://www.buzzfeednews.com/article /craigsilverman/viral-fake-election-news-outperformed-real-news-on-facebook.

211 *Some Facebook executives, like Adam*: Salvador Rodriguez, "Facebook's Adam Mosseri Fought Hard Against Fake News—Now He's Leading Instagram," *CNBC*, May 31, 2019, https://www.cnbc.com/2019/05/31/insta gram-adam-mosseri-must-please-facebook-investors-and-zuckerberg.html.

212 *In their attempt to be impartial*: Sarah Frier, "Trump's Campaign Says It Was Better at Facebook. Facebook Agrees," Bloomberg.com, April 3, 2018, https://www.bloomberg.com/news/articles/2018-04-03/trump-s-campaign -said-it-was-better-at-facebook-facebook-agrees.

213 *They tended to promote her brand*: Frier, "Trump's Campaign Says It Was Better at Facebook."

213 *He warned the CEO that*: Adam Entous, Elizabeth Dwoskin, and Craig

Timberg, "Obama Tried to Give Zuckerberg a Wake-Up Call over Fake News on Facebook," *Washington Post*, September 24, 2017, https://www .washingtonpost.com/business/economy/obama-tried-to-give-zuckerberg-a -wake-up-call-over-fake-news-on-facebook/2017/09/24/15d19b12-ddac-4ad5 -ac6e-ef909e1c1284_story.html.

213 *Zuckerberg reassured the outgoing president*: Entous, Dwoskin, and Timberg, "Obama Tried to Give Zuckerberg a Wake-Up Call."

214 *While people spent an average*: Sarah Frier, "Facebook Watch Isn't Living Up to Its Name," *Bloomberg Businessweek*, January 28, 2019, https://www .bloomberg.com/news/articles/2019-01-28/facebook-watch-struggles-to -deliver-hits-or-advertisers.

216 *As* The Verge *wrote at the time*: Chris Welch, "Facebook Is Testing a Clone of Snapchat Stories Inside Messenger," *The Verge*, September 30, 2016, https:// www.theverge.com/2016/9/30/13123390/facebook-messenger-copying-snapchat.

219 *When Instagram finally talked*: Thompson, "Mr. Nice Guy."

219 *By December 2016, Instagram was letting*: Sara Ashley O'Brien, "Instagram Finally Lets Users Disable Comments," *CNN Business*, December 6, 2016, https://money.cnn.com/2016/12/06/technology/instagram-turn-off-comments /index.html.

221 *Facebook had already funded research*: Eytan Bakshy, Solomon Messing, and Lada A. Adamic, "Exposure to Ideologically Diverse News and Opinion on Facebook," *Science* 348, no. 6239 (June 5, 2015): 1130–32, https://science .sciencemag.org/content/348/6239/1130.abstract.

221 *"In times like these . . . all of us."*: Mark Zuckerberg, "I know a lot of us are thinking . . . ," Facebook, February 16, 2017, https://www.facebook.com/zuck /posts/10154544292806634.

224 *"We have seen no evidence . . . election,"*: Tom LoBianco, "Hill Investigators, Trump Staff Look to Facebook for Critical Answers in Russia Probe," CNN .com, July 20, 2017, https://edition.cnn.com/2017/07/20/politics/facebook -russia-investigation-senate-intelligence-committee/index.html.

225 *"The data on Instagram isn't complete,"*: Sarah Frier, "Instagram Looks Like Facebook's Best Hope," *Bloomberg Businessweek*, April 10, 2018, https://www .bloomberg.com/news/features/2018-04-10/instagram-looks-like-facebook -s-best-hope.

226 *They were posting photos of products*: Sarah Frier, "Instagram Looks Like Facebook's Best Hope."

11 | THE OTHER FAKE NEWS

231 *"It used to be . . . internet."*: Ashton Kutcher, interview with the author, phone, July 9, 2019.

231 *"I choose to comment . . . more users,"*: Bridget Read, "Here's Why You Keep Seeing Certain Instagram Commenters Over Others," *Vogue*, May 4, 2018, https://www.vogue.com/article/how-instagram-comments-work.

233 *In response, those seeking fame*: Emma Grey Ellis, "Welcome to the Age of the Hour-Long YouTube Video," *Wired*, November 12, 2018, https://www.wired.com/story/youtube-video-extra-long/.

235 *"Consumers have the right . . . advertising,"*: Federal Trade Commission, "Lord & Taylor Settles FTC Charges It Deceived Consumers Through Paid Article in an Online Fashion Magazine and Paid Instagram Posts by 50 'Fashion Influencers,'" press release, March 15, 2016, https://www.ftc.gov/news-events/press-releases/2016/03/lord-taylor-settles-ftc-charges-it-deceived-consumers-through.

236 *An early influencer marketing agency*: "93% of Top Celebrity Social Media Endorsements Violate FTC Guidelines," MediaKix, accessed September 20, 2019, https://mediakix.com/blog/celebrity-social-media-endorsements-violate-ftc-instagram/.

239 *After an FBI investigation and a class action*: Lulu Garcia-Navarro and Monika Evstatieva, "Fyre Festival Documentary Shows 'Perception and Reality' of Infamous Concert Flop," NPR.org, January 13, 2019, https://www.npr.org/2019/01/13/684887614/fyre-festival-documentary-shows-perception-and-reality-of-infamous-concert-flop.

240 *Several of them cited a study*: Agam Bansal, Chandan Garg, Abhijith Pakhare, and Samiksha Gupta, "Selfies: A Boon or Bane?," *Journal of Family Medicine and Primary Care* 7, no. 4 (July–August 2018): 828–31, https://www.ncbi.nlm.nih.gov/pmc/articles/PMC6131996/.

241 *The market for travel reached $8.27 trillion*: World Travel and Tourism Council, "Travel & Tourism Continues Strong Growth Above Global GDP," press release, February 27, 2019, https://www.wttc.org/about/media-centre/press-releases/press-releases/2019/travel-tourism-continues-strong-growth-above-global-gdp/.

242 *"Experiences play into . . . shareable stories."*: Dan Goldman, Sophie Marchessou, and Warren Teichner, "Cashing In on the US Experience Economy," McKinsey & Co., December 2017, https://www.mckinsey.com/industries/private-equity-and-principal-investors/our-insights/cashing-in-on-the-us-experience-economy.

242 *In 2018, the number of plane passengers*: "Air Travel by the Numbers," Federal Aviation Agency, June 6, 2019, https://www.faa.gov/air_traffic/by_the_numbers/.

242 *At a selfie factory called Eye Candy*: Lauren O'Neill, "You Can Now Take Fake Private Jet Photos for Instagram in Toronto," *blogTO*, May 2019, https://www.blogto.com/arts/2019/05/photos-fake-private-jet-instagram-toronto/.

243 *And they're not slowing down; they've raised*: Megan Bennett, "No Eternal Return for Small Investors," *Albuquerque Journal*, August 6, 2019, https://www.abqjournal.com/1350602/no-eternal-return-for-small-investors.html.

243 *Facetune was Apple's most popular*: Kaya Yurieff, "The Most Downloaded iOS Apps of 2017," CNN.com, December 7, 2017, https://money.cnn.com/2017/12/07/technology/ios-most-popular-apps-2017/index.html.

243 *"I don't know what real skin looks like anymore,"*: Chrissy Teigen (@chrissyteigen), "I don't know what real skin looks like anymore. Makeup ppl on instagram, please stop with the smoothing (unless it's me) just kidding (I'm torn) ok maybe just chill out a bit. People of social media just know: IT'S FACETUNE, you're beautiful, don't compare yourself to people ok," Twitter, February 12, 2018, 7:16 a.m., https://twitter.com/chrissyteigen/status/962933447902842880.

243 *The worldwide market for Botox injections*: Market Watch, "Botox: World Market Sales, Consumption, Demand and Forecast 2018–2023," press release, December 10, 2018, https://www.marketwatch.com/press-release/botox-world-market-sales-consumption-demand-and-forecast-2018-2023-2018-12-10 (link removed as of November 2019).

244 *"These filters and edits . . . worldwide."*: Susruthi Rajanala, Mayra B. C. Maymone, and Neelam A. Vashi, "Selfies: Living in the Era of Filtered Photographs," *JAMA Facial Plastic Surgery* 20, no. 6 (November/December 2018): 443–44, https://jamanetwork.com/journals/jamafacialplasticsurgery/article-abstract/2688763.

244 *The BBL procedure was performed on more*: Jessica Bursztyntsky, "Instagram Vanity Drives Record Numbers of Brazilian Butt Lifts as Millennials Fuel Plastic Surgery Boom," CNBC.com, March 19, 2019, https://www.cnbc.com/2019/03/19/millennials-fuel-plastic-surgery-boom-record-butt-procedures.html.

245 *In 2017 a task force representing board-certified*: American Society of Plastic Surgeons, "Plastic Surgery Societies Issue Urgent Warning About the Risks Associated with Brazilian Butt Lifts," press release, August 6, 2018, https://www.plasticsurgery.org/news/press-releases/plastic-surgery-societies-issue-urgent-warning-about-the-risks-associated-with-brazilian-butt-lifts.

245 *"Sad news to all . . . so much,"*: Instagress (@instagress), "Sad news to all of you who fell in love with Instagress: by request of Instagram we've closed our web-service that helped you so much," Twitter, April 20, 2017, 12:34 p.m., https://twitter.com/instagress/status/855006699568148480.

247 *At the end of 2017, an investment*: Malak Harb, "For Huda Kattan, Beauty Has Become a Billion-Dollar Business," *Washington Post*, October 14, 2019, https://www.washingtonpost.com/entertainment/celebrities/for-huda-kattan-beauty-has-become-a-billion-dollar-business/2019/10/14/4e620a98-ee46-11e9-bb7e-d2026ee0c199_story.html.

247 *"Who are we? . . . and needs."*: Emily Weiss, "Introducing Glossier," *Into the*

Gloss (blog), Glossier, October 2014, https://intothegloss.com/2014/10/emily
-weiss-glossier/.

248　*In May 2017, in a widely publicized*: Royal Society for Public Health, "Instagram Ranked Worst for Young People's Mental Health," press release, May
19, 2017, https://www.rsph.org.uk/about-us/news/instagram-ranked-worst
-for-young-people-s-mental-health.html.

251　*That month, research groups commissioned*: "The Disinformation Report,"
New Knowledge, December 17, 2018, https://www.newknowledge.com/articles
/the-disinformation-report/.

12 | THE CEO

253　*In 2012, when Facebook reached*: Leena Rao, "Facebook Will Grow Head
Count Quickly in 2013 to Develop Money-Making Products, Total Expenses
Will Jump by 50 Percent," *TechCrunch*, January 30, 2013, https://techcrunch
.com/2013/01/30/zuck-facebook-will-grow-headcount-quickly-in-2013-to
-develop-future-money-making-products/.

255　*The new hierarchy, the biggest reshuffling*: Kurt Wagner, "Facebook Is Making Its Biggest Executive Reshuffle in Company History," *Vox*, May 8, 2018,
https://www.vox.com/2018/5/8/17330226/facebook-reorg-mark-zuckerberg
-whatsapp-messenger-ceo-blockchain.

256　*Whatever they wanted to do was their right*: Parmy Olson, "Exclusive:
WhatsApp Cofounder Brian Acton Gives the Inside Story on #DeleteFace
book and Why He Left $850 Million Behind," *Forbes*, September 26, 2018,
https://www.forbes.com/sites/parmyolson/2018/09/26/exclusive-whatsapp
-cofounder-brian-acton-gives-the-inside-story-on-deletefacebook-and-why
-he-left-850-million-behind/.

256　*The consensus was that the team had*: Kirsten Grind and Deepa Seetharaman,
"Behind the Messy, Expensive Split Between Facebook and WhatsApp's
Founders," *Wall Street Journal*, June 5, 2018, https://www.wsj.com/articles
/behind-the-messy-expensive-split-between-facebook-and-whatsapps
-founders-1528208641.

257　*"I find attacking the people . . . low-class,"*: David Marcus, "The Other Side of
the Story," Facebook, September 26, 2018, https://www.facebook.com/notes
/david-marcus/the-other-side-of-the-story/10157815319244148/.

258　*On Friday, March 17, 2018, the* New York Times: Matthew Rosenberg, Nicholas Confessore, and Carole Cadwalladr, "How Trump Consultants Exploited
the Facebook Data of Millions," *New York Times*, March 17, 2018, https://
www.nytimes.com/2018/03/17/us/politics/cambridge-analytica-trump
-campaign.html; and Carole Cadwalladr and Emma Graham-Harri-
son, "Revealed: 50 Million Facebook Profiles Harvested for Cambridge

Analytica in Major Data Breach," *The Observer*, March 17, 2018, https://www
.theguardian.com/news/2018/mar/17/cambridge-analytica-facebook
-influence-us-election.

261 *The average Facebook employee made*: Casey Newton, "The Trauma Floor,"
The Verge, February 25, 2019, https://www.theverge.com/2019/2/25/18229714
/cognizant-facebook-content-moderator-interviews-trauma-working
-conditions-arizona; and Munsif Vengatil and Paresh Dave, "Facebook Contractor Hikes Pay for Indian Content Reviewers," Reuters, August 19, 2019,
https://www.reuters.com/article/us-facebook-reviewers-wages/facebook
-contractor-hikes-pay-for-indian-content-reviewers-idUSKCN1V91FK.

261 BuzzFeed News *tallied that between*: Alex Kantrowitz, "Violence on Facebook
Live Is Worse Than You Thought," *BuzzFeed News*, June 16, 2017, https://
www.buzzfeednews.com/article/alexkantrowitz/heres-how-bad-facebook
-lives-violence-problem-is.

262 *Deaths from opioids had more than doubled*: "Overdose Death Rates," National Institute on Drug Abuse, January 2019, https://www.drugabuse.gov
/related-topics/trends-statistics/overdose-death-rates.

262 *"Yikes," Rosen responded*: Sarah Frier, "Facebook's Crisis Management Algorithms Run on Outrage," *Bloomberg Business*, March 14, 2019, https://
www.bloomberg.com/features/2019-facebook-neverending-crisis/.

273 *"Instagram's Co-Founders to Step Down from Company."*: Mike Isaac, "Instagram Co-Founders to Step Down from Company," *New York Times*,
September 24, 2018, https://www.nytimes.com/2018/09/24/technology/instagram
-cofounders-resign.html.

273 *"Mike and I are grateful . . . companies do next."*: Kevin Systrom, "Statement
from Kevin Systrom, Instagram Co-Founder and CEO," Instagram-press.com,
September 24, 2018, https://instagram-press.com/blog/2018/09/24/statement
-from-kevin-systrom-instagram-co-founder-and-ceo/.

274 *But because of the leaks to the media*: Sarah Frier, "Instagram Founders Depart
Facebook After Clashes with Zuckerberg," Bloomberg.com, last modified
September 25, 2018, https://www.bloomberg.com/news/articles/2018-09-25
/instagram-founders-depart-facebook-after-clashes-with-zuckerberg.

274 *The press was writing about mounting tensions*: Frier, "Instagram Founders
Depart Facebook."

EPILOGUE

279 *In 2019, Instagram delivered*: Sarah Frier and Nico Grant, "Instagram Brings In
More Than a Quarter of Facebook Sales," Bloomberg.com, February 4, 2020,
https://www.bloomberg.com/news/articles/2020-02-04/instagram-generates
-more-than-a-quarter-of-facebook-s-sales.

INDEX

ABC News, 211

Above Category cycling, 185

abuse, abuse content, 41, 97, 261

Academy Awards, 152
   Systrom at, 191–92, 204

Accenture, 260

Acton, Brian, 125, 256, 258

Adams, William (will.i.am), 128

Adidas, xix

Adobe Lightroom, 240, 243

Adobe Photoshop, 21, 23, 244

advertising, 59, 176, 256
   false, 244; *see also* fake news
   FB's business of, 75, 77, 89, 91–92,
      94, 96, 105, 118–19, 125, 149–50,
      163, 217, 224, 277
   IG's business of, 104, 118–21, 124,
      151, 155, 163–65, 174, 175–76,
      184, 225, 241, 277
   mobile, 74–75
   television, 215
   see also brand advertising

advertising agencies, 89
   FB's relationship with, 120–21,
      124

Ahrendts, Angela, 147

@aidanalexander, 171

AiGrow, 246

Airbnb, xvi, 45

Alba, Jessica, 130, 250

Alexander, Aidan, 171

algorithms:
   FB's use of, 91, 103, 128, 162, 163,
      208, 209, 210–12, 215, 221, 224,
      259
   IG's early lack of, 34, 143
   IG's use of, 81, 170, 174, 197–98,
      218, 229, 230–32, 233, 251, 271
   IG users' mistrust of, 197
   YouTube's use of, 233–34

@alittlepieceofinsane, 161

Allen, Nick, 117

Allen & Company, 49

Amanpour, Christiane, 127

Amaro photo filter, 23

Amazon, 22, 28, 139, 242
   Communications Decency Act and,
      41
   Whole Foods acquired by, 64
   Zappos acquired by, 105

Amazon Web Services, 26, 79–80

American Medical Association, 244

American Society of Plastic Surgeons,
   244

analytics, 90, 102, 226
  IG's use of, 100, 178, 183, 226
  IG users' access to, 275–76
Anchor Psychology, 172
Anderson, Steve, 34
  early IG investment of, 11, 15
  on IG board, 37, 56, 63–64
Anderson Cooper 360, 142
Andreessen, Marc, 11
Andreessen Horowitz:
  early investment in IG by, 11, 15, 33
  investment in PicPlz by, 33–34, 36, 77
Android, 19, 33, 110, 203
  IG app for, 50, 51
angel investments, 16, 17, 24, 36
anonymity, user, 41, 80
  on Formspring, 40
  on IG, 41, 80, 163, 173, 218, 219, 260, 261
  troubling content and, 40, 163, 218–19, 260; *see also* bullying
antitrust laws, 75, 268
Antonow, Eric, 165
AOL, 116
Apple, 10, 21, 56, 65, 147, 167, 234
Apple app store, 26, 28, 38, 115
Apple IDs, 42
apps:
  filter, *see* filter apps, photo
  location-based, 15
  see also mobile apps; *specific apps*
Argentina, 12
Arthur D. Little, 241
Atlantic, 68
Automattic, 17
Axente, Vanessa, 156

Bach, King, 110
@backpackdiariez, 240

Bailey, Christopher, 145–47
Bailey, Will, 199
Bain, Adam, 194
Bali, 242
Barbie doll, 136
Barbour, 167
BarkBox, 142
Barnett, John, 187–88, 190, 192, 193, 199, 202
Barnieh, Edward, 167, 168–69, 246
Baseline Ventures, 11, 15
Beast (Hungarian sheepdog), 60, 61–62, 67
Beck, Glenn, 208
Beco do Batman (Batman's Alley, São Paulo), xv–xvi
Ben & Jerry's, 119
Benchmark Capital, 36–37, 46, 55, 56
#benoteworthy, 82
beta-testing, 22, 24, 219
#betterworld, 82
Beyoncé, 230
Bezos, Jeff, 22, 105
Bickert, Monika, 225
Bieber, Justin, 44, 48, 51, 98, 131, 174
  as early IG user, 38–39
Bikini Blast, 10
bikini shots, 170
Bilton, Nick, 14, 98, 99
#binghazi, 181
Birdman (film), 158
Black Eyed Peas, 128
@blackstagram, 251
Black Tap, xviii
#blessed, 230
Blink-182, 238
Blogger, 6
blogs, bloggers, 136, 143, 155, 237, 246, 265
Bloomberg Businessweek, 175
Bloomberg News, 213

Blue Bottle Coffee, 108, 205
Bono, 127
#bookstagram, 171
Bosstick, Lauryn Evarts, 237
Bosworth, Andrew "Boz," 94–95, 105, 107, 164
Botha, Roelof, 55
Botox injections, 243
bots, 173–75, 245–46, 259
Bourgeois, Liz, 154–55
brand advertising, by IG users, xvii, xviii, 82–83, 104, 118–21, 139, 153, 155, 219, 229–30, 235–37, 245, 246–47
    failure to disclose income from, 35–36, 138, 235–37
    influencers and, 138–39, 167, 235–36
    see also advertising
brands, branding, 35
    of FB, 94, 100, 209, 216, 222, 266
    of IG, 27, 41, 89, 94, 100, 104, 111, 119, 132, 160, 162, 164, 177, 209, 216, 217–18, 254
    personal, xvi–xvii, xx, 113, 140, 189, 237
Branson, Rick, 79, 80
Braun, Scooter, 39, 48
Brazil, 12, 184, 203
Brazilian butt lift, 244–45
Breitbart, 208
Brenner, Kevin, 243–45
Brexit, 220
BroBible, 113
Brown, Reggie, 113
Brunelleschi, Filippo, 3
Bublé, Michael, 129
Budweiser, 232
bugs, bug fixes, 38, 110
bulimia, 41
Bull, Janelle, 172, 173

bullying, 40, 41, 135, 161, 163, 218–19, 248, 271
Burberry, 145–47
Burbn, 10–11, 18, 101
    early investment in, 11, 15, 16
    iPhone app version of, 17
    square photo format of, 19
    Systrom's doubts about, 16–17
    see also Instagram
Bureau of Consumer Protection, 235
Burstn, 33
Business Insider, 68
BuzzFeed News, 261

California Department of Corporations, 99
    IG acquisition "fairness hearing" of, 84–86, 98
Calvin Klein, xix
Camarena, Meghan, 171
Cambridge Analytica, 258, 259
Camera+, 21, 76
Camera Awesome, 76
Campbell, Edie, 156
Campbell, Naomi, 236
#campvibes, 229
Captiv8, 238
Carey, Eileen, 261–62, 278
Carlson, Tucker, 208
#catband, 155
Catholic Church, 196, 204
celebrity, celebrities, 138, 184, 254, 275
    brand advertising on IG by, 155, 232, 236
    FB and, 126–28, 130, 149–50, 156, 216
    IG-created, xvii–xviii, xix, 130, 139, 152, 171
    IG partnerships with, 25, 128, 160–61, 218–19, 230, 235, 250, 264, 279

celebrity, celebrities (*cont.*)
  IG Stories and, 203–4
  IG use by, 35–36, 40, 46–48, 83, 128,
    134, 140, 147, 149–50, 184, 231
  Snapchat use by, 192, 204
  Twitter use by, 7, 38–39, 46, 130, 156
  wooed by IG, 128–29, 131–36, 139,
    148, 153, 155, 171, 264
  YouTube use by, 130
cell phones, *see* mobile phones
Cerny, Amanda, 112
Chafkin, Max, 175
Chan, Priscilla, 221
Charlie (photography teacher), 5, 8,
  19, 21
Chen, Pamela, 153, 204
child pornography, 41, 42
Choi, Christine, 190, 193
Christensen, Clayton, 267–68
Chrysler, 47
Ciara, 236
Cicilline, David, 69
Clinton, Hillary, 181, 210, 212, 251
  FB and, 212–13
  Systrom and, 207–8
CNET, 99
CNN, 54, 127, 148, 224
  IG account of, 35, 44
Coca-Cola, 118
#cocaine, 262
Codename, 20–21, 23, 24
  see also Instagram
Coffee Bar, 12
Cognizant, 260
Cohler, Matt, 36–37, 38
  at Facebook, 37, 64
  on IG board, 37, 56, 60, 64
Colao, J. J., 113
Cole, Amy, 47, 53, 73, 105, 131
comments, xvii, 31, 144, 157, 158, 175,
  233, 236, 251, 271

@commentsbycelebs, 233
Communications Decency Act (1996),
  41–42
community, communities, 260
  global, 221, 224, 268
  at IG, 21, 168
  interest-based, xvii, 46, 153–54, 171
  social network, 19, 109
  valued at IG, 34, 40, 72, 94, 95,
    102, 108–9, 139, 147, 205, 271,
    272, 274; *see also* Instagram,
    community team at; InstaMeets
Congress, U.S., 41, 264, 278
  Zuckerberg's testimony to, 258, 262
  see also House of Representatives,
    U.S.; Senate, U.S.
content, sponsored, *see* brand
  advertising; brands, branding
content, user, xix, 36, 39, 81, 92, 100,
  134, 143, 149, 171, 184, 196–97,
  210, 219–20, 225, 233–34, 246,
  270, 276, 278
  ephemeral, 113–14, 117, 190–92;
    see also Instagram Stories;
    Snapchat
  fake, *see* fake news
  FB and, 149, 156, 207–8, 211, 212,
    215, 216, 259
  "Instagrammable," 172
  original vs. re-shared, 157, 269
  pressure to produce, 237, 241–42
  re-sharing of, disallowed on IG, 20,
    43–44, 45, 140
  troubling and abusive, 41–43, 80,
    97, 225, 249, 260–62, 271
  virality and, 162, 260
  see also Instagram, content curated
    by
Conway, Ron (father), 17
Conway, Ronny (son), 15, 17–18
Cooper, Anderson, 142

Cooper, Bradley, 152

Cooper, Sia, 231

Costolo, Dick, 46–47, 150, 151
  in Twitter's attempt to buy IG, 49, 55

Cox, Chris, 103, 107, 225, 254, 255, 266
  Systrom and, 257, 267–68, 272

Craigslist, 129

creators, celebrity, 219

Crime Desk SF, 12, 32

cross-processing, 21

Crowley, Dennis, 28–29

Cyrus, Miley, 131, 160–61, 162–63, 218

daily active users (DAUs), 93

Daily Mail, 68

D'Angelo, Adam, 24, 26, 36

@danrubin, 30

@darcytheflyinghedgehog, 48

Dash, 137

Dasher, Courtney, 141–42, 153

Daum Kakao, 77

Deadmau5 (Joel Thomas
  Zimmerman), 70

DeGeneres, Ellen, 152

#deletefacebook, 258

#deleteinstagram, 99

Delettrez-Fendi, Delfina, 195

Delevingne, Cara, 156

Demyttenaere, Camille, 239–41

Deschanel, Zooey, 130

#designlab, 235

Develin, Mike, 183–84, 226–27

Diaconu, Andreea, 156

Diamond, Emma, 232

@diaryofafitmommyofficial, 231

Dishd, 10

Disney, Walt, 219

Disneyland, 219

Dixon, Chris, 11

DJ Khaled, 171

Dogpatch Labs, 17, 18, 23, 32

Doom II, 4

dopamine, xxi

Dorsey, Jack, 83–84, 98, 109, 194
  betrayal felt by, 65–66, 194–95
  first post to IG by, 25
  investment in IG by, 16, 36
  at Odeo, 6, 7, 27
  Systrom and, 6–7, 15–16, 19, 24, 60,
    66, 84
  at Twitter, 14, 46–47
  and Twitter's attempt to buy IG, 46,
    49, 55
  Williams and, 25–26

dot-com boom and bust, 5, 8

#doubleinsta, 184

Dovetale, xviii, 238

Doww, Jordan, 171

Dude, Where's My Car? (film), 44

Dunham, Lena, 186

DuShane, Renee, 161

Earlybird photo filter, 22

eBay, 36, 110

Ebersman, David, 58

Eckert, Emily, 222

elections, U.S., of 2016, 207–8, 209,
    210–11, 214, 220, 226, 232
  fake news in, 211–13, 234, 254, 261
  FB's effort to appear equitable in,
    208
  Russian interference in, 224–25,
    235, 255, 271
  see also Clinton, Hillary; Trump,
    Donald

elections, U.S., of 2020, 278

1197 conference, 40

emojis, 39, 218

E! News, 154, 192

escapism, 23, 209, 217, 239, 241

Escobar, Pablo, 238

Eswein, Liz, 43–44, 48, 82
Everson, Carolyn, 120–21
Eye Candy, 242

Facebook, xviii, 7, 10, 19, 39, 111, 181,
    217, 222, 227, 246, 248, 253
  advertising agencies' relationship
      with, 120–21, 124
  advertising business of, 75, 77,
      91–92, 94, 96, 105, 118–19, 125,
      149–50, 163, 217, 224
  algorithmic personalization
      approach of, 91, 103, 128, 162,
      163, 208, 209, 210–12, 215, 221,
      224, 259
  board of, 57, 63, 125, 191
  business model of, 258–59
  business team at, 222
  Cambridge Analytica scandal at,
      258, 259, 267
  celebrity outreach of, 127–28, 156
  Communications Decency Act and,
      41
  communications team at, 211, 222
  "connect the world" goal of, 91,
      149, 162
  content policing at, 43, 97, 259,
      260–61
  Creative Labs at, 124
  Creative Labs Skunk Works at, 191
  cryptocurrency initiative at, 257,
      263
  data collection by, 89, 90, 91–92, 93,
      122, 125, 143, 149, 258–59
  dominance of, in social network
      world, 78, 88, 121, 124, 151, 209,
      253, 255
  early IG acquisition interest of, 28
  edge stories of, 209–10
  employee handbook of, 65, 93, 106,
      228

  engineering team at, 60, 62
  ephemeral sharing on, 191, 193,
      214, 215–16, 217
  fake news scandal at, 210–11,
      224–25, 234, 250–51, 255
  "family of apps" bundled together
      with, 255–56, 259, 267–68, 277,
      279
  as friend-based network, 20, 31, 80
  "friend coefficient" of, 91
  global partnerships team at, 148
  Gowalla acquired by, 51
  growth as overarching goal at, xvii,
      9, 91–92, 96, 150–51, 160, 209,
      276
  growth team at, 90, 92, 93, 95, 122,
      177, 223, 260, 268
  hackathons hosted by, 124, 187
  hacker culture at, 93, 102
  hierarchy reshuffle at, 255–56
  hyperlinks on, 80, 210
  as ideological echo chamber,
      220–21
  IG photo sharing to, 37
  IG's access to infrastructure and
      resources at, 96, 159, 162, 225,
      249–50
  IG's access to infrastructure and
      resources denied by, 262, 268–69
  IG's cannibalization of, as issue at,
      223, 226, 227–28, 257, 280
  IG seen as threat to, xvii, 38, 57,
      63–64, 77–78, 90, 95, 252–53,
      270
  IG's growing resentment of, 254,
      262, 263, 274
  IG's independence at, 54, 63, 65,
      67, 89, 96, 106, 118, 121, 124,
      209, 222–23
  IG's integration at, 100–101, 114,
      223

IG users' mistrust of, 100, 197
integrity team at, 259–60, 271
Internet.org division of, 124
IPO of, 56, 58, 74, 84, 128, 150
leadership team at, 222
link added from IG to, 228, 257
link to IG removed from, 228, 269
live video on, 261
lockdowns at, 108, 256, 269–70
Lookalike Audience tool of, 213
Menlo Park headquarters of, 66–68,
    179, 200, 203, 210, 254
Mentions app of, 156
mobile advertising of, 105, 150
mobile troubles of, 93, 98
Mobile Uploads album on, 28
mounting tensions between IG and,
    262, 263, 274
network effects of, 77–78
news feed on, 81, 91, 94, 96, 163,
    211–12, 215, 259, 263, 264, 269
Nextstop acquired by, 32
Onavo acquired by, 122, 123, 181
1-billion-user milestone reached by,
    88, 265
operations team at, 225
Palo Alto office of, 15, 88, 107
Paper app of, 156
partnerships team at, 50, 148, 200
photos launched by, 7
Poke app of, 115–16, 124, 191
policy team at, 211, 225, 232
politics and, 209–13
privacy issues at, 54, 92, 94, 264
privacy scandals at, 28, 154, 255,
    267
public policy team at, 211
Riffs app of, 191
Saverin and, 15
scale and, 98, 103, 143, 225, 265
Slingshot app of, 191

tagging friends on, 7, 90
technology team at, 24, 26, 95, 214
trending topics module of, 207,
    210
2-billion-user milestone reached
    by, 252
video on, 109, 215, 253–54
virality on, 162, 209, 211, 215, 251,
    260
WhatsApp acquired by, 125, 202,
    255
Yahoo! in attempted purchase of, 57
youth team at, 202
and Zuckerberg's attempts to buy
    Snapchat, 114–15, 116, 117, 122,
    125, 183, 191, 200–202
Zuckerberg's unilateral power at,
    63, 116, 202
Facebook, Instagram acquired by, xvii,
    xx, 52–53, 56–65, 92, 255, 278
California Department of
    Corporations "fairness hearing"
    on, 84–86, 98
FTC approval of, 71, 78, 84
FTC investigation of, 74, 75–76,
    78
IG employees as financial losers in,
    71–73
$100 billion valuation of IG in, 58,
    61, 74, 77
ripple effect on industry of, 109–10,
    114
Systrom and Krieger made wealthy
    by, 72
Systrom's negotiations with
    Zuckerberg in, 57–58, 60–62, 73
Facebook Album, 28
Facebook Camera, 76, 90
Facebook Messenger, 120, 214, 216,
    255, 257, 266, 268, 276
Facebook Stories, 214, 216–17, 256

Facebook users, 38, 54, 56, 59, 74, 75, 77, 78, 88, 90–91, 96, 124, 126, 149, 150, 162, 201, 210–11, 212, 222–24, 253, 269
  as DAUs, 93
  FB's collection of data of, 89, 90, 91–92, 93, 122, 125, 143, 149, 154, 258–59
  friend networks of, 154, 209
  privacy issues and, 94, 258–59
  real identity of, required by FB, 80, 173, 260
  time spent per day on FB, 91–92, 214–16, 228, 235, 248, 276
  virality on, 162, 209, 211, 215, 251, 260
  Zuckerberg's resolve to better understand, 221–22
  see also content, user
Facebook Watch, 215, 254–55, 265
Facetune, 243
Fake, Caterina, 11
fake news, 213, 224, 266
  FB's algorithms gamed by, 210–11, 234
  IG gamed by, 251
  propagated by Russia, 224–25, 235, 250, 255
"famous for being famous" idea, 136
fashion community, 145–47, 264
  IG and, 131–32, 156, 195
Federal Bureau of Investigation (FBI), 239
Federal Trade Commission (FTC), U.S., 69, 94, 235–36, 255, 278
  Bureau of Consumer Protection of, 235
  FB acquisition of IG approved by, 71, 78, 84
  FB acquisition of IG investigated by, 74, 75–76, 78–79

Fendi, Silvia Venturini, 195
Fenton, Peter, 46, 55
Fenty Beauty, 232
Fenwick & West, 76
FilmLoop, 5
filter apps, photo, xxi, 21, 22, 23, 29, 31, 34, 35, 45, 47, 75, 76, 114, 147, 167, 173, 245
financial crisis of 2008–2009, 9, 35
finsta IG accounts, 182–83, 184, 243
Flickr, 11, 24, 76
  acquired by Yahoo!, 54
Flipboard, 156
Florence, Italy, Systrom's study in, 3, 4–5, 19, 21
followers, following, xvii–xviii, xxi, 20, 31, 139–40, 172, 173, 233, 234, 236, 279
  reciprocal, 183–84
  on Twitter, 25, 26, 34, 36, 44, 84
  on Vine, 110, 111, 112, 157
followers, following, on IG, xvii–xx, 26, 27, 35, 36, 44, 59, 80, 82, 108–9, 146, 163, 233–34, 236, 237, 268, 270, 276, 279
  brand advertising and, xviii, 119, 246–47; *see also* brand advertising, of IG users
  of celebrities, 35–36, 38–39, 131, 134–36, 138–39, 160–61, 192, 203–4, 218–19, 230–32, 265
  fake, 173–75
  gaming the system in acquiring of, 144, 168–69, 173, 231, 245–46
  hashtags and, 140
  and IG's feed order algorithm, 197, 229–30
  influencers and, 231, 237–38
  @instagram account and, 103, 141–42, 167, 171, 203–4
  media and, 155

Popular page and, 144
and pressure for good content,
  241–42
risky behavior in acquiring of,
  169–70, 239–40
suggested user list and, 48, 103, 141,
  142–43, 270, 277
teens' obsession with, 172, 182–84,
  192
#followmebro, 169
Food and Drug Administration
  (FDA), 36
"food porn," 81
Forbes, xx, 113, 114, 200, 256
Ford Motor Co., 118
Formspring, 40, 42
Formula One, 204
FOSes (friends of Sandberg), 101
Foster, David, 129
Foster, Erin, 129–30
Foursquare, 14, 15, 19, 28, 29, 55
Fox Interactive Media, 24
Fox News, 208
Francis, Pope, 195–96
free speech, 232–33
#fromwhereirun, 163
Froome, Chris, 148
FuckJerry, 239
Fulk, Ken, 185, 186
Furlan, Brittany, 110, 112
Fyre Festival scandal, 238–39

Gabriel (Brazilian IG user), xvi
Game of Thrones (TV series), 64
Gant boutique, 95
Gates, Bill, 200
@gdax, 48
Gehry, Frank, 115
Gil, Elad, 48, 49
Gizmodo, 37, 207–8
Glass, Ira, 188–90

Glossier, 247–48
Gmail, 8–9, 24
Golden Globes, 127, 192
Gomez, Selena, 131, 230
  as early IG user, 39
González Iñarrítu, Alejandro, 158
Google, 8–9, 56, 59, 202, 225, 234
  Blogger acquired by, 6
  early IG acquisition interest of, 28
  Nest acquired by, 64
  social network launched by, 108
  Systrom at, 7, 8–9, 11, 23, 37, 62,
    194
  YouTube acquired by, 53, 59, 105
Google search, 155, 183
Gorgon, Cecilia, 247
Gowalla, 14, 15, 51, 53, 73
Grammy Awards, 36
Grande, Ariana, 134–35, 142, 218, 230
Great Recession of 2008–2009, 9, 35,
  53
Grier, Nash, 110
Griswold, Ivan, 86
Groban, Josh, 129
growth hacking, 231
Guardian, 99, 166
Gucci, 195
Gulf of Mexico, BP oil spill in, 113
Gutierrez, Manny, 265

Hadid, Bella, 238
Hamilton, Lewis, 204
Hammam, Imaan, 156
Hansen, Scott (Tycho), 34–35
Happy Hippie Foundation, 160
Harris, Calvin, 218
Harvard University, 1
Harvey, Del (pseud.), 42
hashtags, xxi, 59, 140, 147, 154–55, 260,
  262, 262, 270
Hasselblad camera, 22

Hatch, Orrin, 259
Hatching Twitter (Bilton), 14
Hathaway, Paige, 144–45, 153
Hensley, Dustin, 243
Hewlett-Packard, 2
Hilton, Paris, 136–37, 139
Hipstamatic, 21, 22, 76
Hochmuth, Gregor, 10, 11, 20–21, 47, 50–51, 52, 90
Hocke, Jean, 239–41
Holga camera, 5
Hollande, François, 186
Holliston, Mass., Systrom's childhood in, 3–4
HomeGoods, 4
Honan, Mat, 37
Hong Kong, InstaMeets in, 168–69, 246
Horsley, Hunter, 176
House of Representatives, U.S., Zuckerberg's testimony to, 258
see also Congress, U.S., Senate, U.S.
Hsieh, Tony, 105
HTML5, 19
@hudabeauty, 247
Hudgens, Vanessa, 236
Hudson, Kate, 192
Hudson photo filter, 23
Huffine, Candice, 250
Huffington, Arianna, 171
Huffington Post, 154, 170
Hughes, Chris, 78
Hunch, 11
hyperlinks, xxi, 8, 80, 210

#iammorethan, 161
Iceland, 242
identity theft, 97
iFart app, 10
@ileosheng, 161

#imwithher, 208
Incredibles, The (film), 180
India, IG Stories and, 203
Indonesia, IG in, 226
influencer economy, xxi, 128–29, 170–71
influencers, 25, 36, 83, 127, 165, 166, 170, 184, 231, 237–38, 240–41, 265
  branding and, 138–39, 167, 235–36; *see also* brand advertising, of IG users
  celebrity, 138–39, 172, 239
  fake followers and, 173–75
  IG analytics increasingly available to, 275–76
  IG as, *see* @instagram
  IG's comments ordering algorithm and, 230–31
  IG's feed order algorithm and, 197–98, 229–30
  pods joined by, 246
  teen digital-first, 171
Insta-bae (Instagrammable design movement), xviii
"INSTAGIRLS, THE" (*Vogue* cover headline), 156, 167
@instagram, xxii, 102, 104, 141, 143, 160–61, 167, 169, 171, 203, 204, 210, 216, 241, 247
Instagram:
  advertising business of, 104, 118–21, 124, 151, 155, 163–65, 174, 175–76, 184, 225, 241, 277
  algorithmic ordering of comments on, 230–32, 233, 251
  algorithmic post order shift by, 197–98, 218, 229
  ambiguous advertising on, 35–36
  ambiguous content rules of, 143
  analytics team at, 183, 226

Android app of, 50, 51

blocked from Twitter access, 84, 99

board of, 37, 56, 60, 62, 63

Boomerang and, 187, 190

bot detection algorithm of, 174

brand advertising on, *see* brand
    advertising

brand of, 27, 41, 89, 94, 100, 104,
    111, 119, 132, 160, 162, 164, 177,
    209, 216, 217–18, 254

bullying on, 41, 135, 161, 163,
    218–19, 271, 279

business model lacked by, 54, 75,
    77, 100, 118, 124–25

business team of, 118

celebrities courted by, 128–29,
    132–36, 139, 264; *see also*
    celebrity, celebrities

as celebrity-making machine,
    xvii–xviii; *see also* influencers

chronological order of posts on,
    117, 196–97

communications team at, 154, 202,
    222, 271

community team at, 80, 81, 103,
    104, 140, 141, 155–56, 160, 166,
    169, 170, 176, 203–4, 226, 234

community valued by, 34, 40, 72,
    94, 95, 102, 108–9, 139, 147, 205,
    271, 272, 274

content curated by, 25, 41, 43, 81,
    103–4, 114, 140–41, 143, 151,
    152, 161–62, 169, 170, 210, 235,
    279; *see also* @instagram

content moderation of, transitioned
    to FB, 97, 225, 249, 260

creativity, design, and experiences
    as focus of, xxi, 35, 66, 83, 91,
    93, 100, 103, 108–9, 128, 139,
    160, 167, 175, 180, 205, 264,
    276

customer service lacked by, 32, 132,
    230

daily habit strategy of, 13–14

direct messaging of, 123, 276

Dorsey's early promotion of, 25–26

earliest incarnations of, *see* Burbn;
    Codename

early investors in, 24, 26, 27, 37

events team at, 265

Explore page on, 170

fake news and, 225, 256

fakery detection algorithm of, 174

fashion industry and, 131–32,
    145–47

FB infrastructure and resources
    available to, 96, 159, 162, 225,
    249–50

FB infrastructure and resources
    denied to, 262, 268–69

as feel-good app, 154, 157, 171,
    210, 217–18, 219–20

finsta accounts on, 182–83, 184,
    243

first photo posted on, 21, 180

first users chosen carefully by, 25,
    33, 34

founders of, *see* Krieger, Mike;
    Systrom, Kevin

Fyre Festival scandal and, 238–39

growing resentment of FB at, 254,
    262, 263, 274

growth rate of, 216

growth team of, 177–78, 269–70

hashtag tool on, xxi, 59, 140, 147,
    154–55, 260, 262, 270

hyperlinks not allowed in, xxi, 80,
    210

IGTV of, 252, 254–55, 257, 264–67,
    270

illegal drug sales content on,
    261–62, 270, 271, 278

Instagram (*cont.*)

independence of, at FB, 54, 63, 65, 67, 89, 96, 106, 118, 121, 124, 209, 222–23

integration of, at FB, 100–101, 114, 223

as internet's utopia, 220, 225

Krieger's resignation from, xxii, 272–75

link to, removed from FB, 228, 269

link to FB added on, 228, 257

live video on, 261

lockdown at, 269–70

logo of, xvi, 20, 34

mainstreaming of, 35, 47, 168, 169, 170, 173

media outreach of, 154–55

mission statement of, 102

as mobile-only app, 27, 89

mounting tensions between FB and, 262, 263, 274

in move out of FB headquarters, 204–5

naming of, 24

network effects of, 77–78

$1 billion revenue milestone reached by, 186

$1 billion valuation of, xx, 53, 54, 58, 85

1-billion-user milestone reached by, 264–65, 267, 280

operations team at, 180, 204

Paradigm Shift program of, 184, 186, 187, 190

partnerships team at, 160, 219, 230, 235

photo filters on, *see* filter apps, photo

photo tagging on, 95

Pixel Cloud of, 190

politics and, 207–8

Popular page of, 81, 140, 144, 170

public policy team at, 160, 249

rebranded as "Instagram from Facebook," 276

reciprocal follower problem at, 183–84

rectangular photo format added on, 176–77

research team at, 199

re-sharing not allowed on, 43, 44, 140, 157

sales team at, 165

seen as threat to FB, xvii, 38, 57, 77–78, 90, 95, 252–53

server meltdowns of, 26, 30, 32, 38–39, 51, 79–81

simplicity valued at, 27, 30, 65, 102–3, 125, 160, 178, 180, 199

Snapchat as threat to, 123, 178, 181, 184, 192–93, 201

sold by Systrom and Krieger to Facebook, *see* Facebook, Instagram acquired by

South Park office of, 32, 44, 52, 79, 181, 193

spam on, 80, 226

square photo format of, 19, 31, 110, 147, 176

suggested user list of, 48, 81, 82, 103, 143, 153, 168

suicide content on, 41–42, 261, 277–78

Systrom as public face of, 33

Systrom's resignation from, xxii, 272–75

teens team at, 161, 170

"terms of service" debacle at, 99–100

Third Thursday Teens series of, 182, 183

translated into other languages, 43, 97

travel influenced by, 169, 241, 242

troubling user content on, 80, 160–61, 260–61, 270

and Twitter's efforts to buy, 25, 46, 48–49, 55–56, 86, 109

2010 launch of, 26–27, 31–32

underlying culture of, 248–49

user anonymity on, 41, 80, 163, 173, 218, 219, 260, 261

user guidelines of, 155

user types cultivated by, 153

verified accounts on, 132–33, 231, 232, 279

video launch of, 110–11, 118, 145

weekend hashtag project of, 104

well-being team at, 249, 260, 271, 275

worldwide impact of, xvi–xx

Zuckerberg's concern about cannibalization of FB by, 223, 226, 227–28, 257, 280

Zuckerberg as taking credit for success of, 266–67

"Instagrammable," xix, 81, 166–67, 172–73, 254, 265–66

Instagrammable design movement, xviii, 168

Instagram Stories, 198–99, 201–4, 205, 207, 214, 226, 227, 245, 248, 250–51, 264

Instagram users, 197, 233

brand advertising by, *see* brand advertising, by IG users

changing behavior of, in posting to IG, 80–81, 83, 169, 172, 233, 239–40, 243

as concerned over FB's acquisition of IG, 54

feelings of inadequacy among, 275

growth hacking by, 231

IG's analytics tools available to, 275

IG's relationships with, 166

IG used as publisher by, 237

pods used by, 246

as pressured to post the best, 29, 114, 172, 173, 175, 178, 275

self promotion by, 233

as unofficial ambassadors for IG, 43–44

see also celebrity, celebrities; content, user; influencers

Instagress, 175, 245–46

InstaMeets, 81, 102

organized by IG, 34–35, 39–40

organized by IG users, 43, 44, 48, 104, 143, 148, 167, 168, 246

instant messaging services, 12

Instazood, 175, 246

Intel, 2

internet, 9, 16, 31, 56, 65, 79–80, 108, 109, 115, 126, 136, 229, 233

early days of, 3–4

FB as largest network on, 78, 88, 163, 253, 255

first generation of, 5

IG as top pop culture destination on, 126, 195

Section 230 of the Communications Decency Act and, 41–42

smartphones and, 10, 40

Web 2.0 and, 5

world population connected to, 124, 163, 234

Internet.org, 124

internet trolls, 219

investors, angel, 16, 17, 24

iPhones, 10, 30–31, 145

Burbn app for, 17

5S launch, 146

IG featured in launches of, 28

Krieger's early apps for, 12

iPhones (*cont.*)
- photo filters on, 147
- photo technology of, 18, 152
- square photo format on, 147, 176

Iribe, Brendan, 217, 253
iTunes, 137

Jackson Colaço, Nicky, 160, 207, 208, 219, 220, 225–26, 249
JAMA Facial Plastic Surgery, 244
James, LeBron, 131
Japan, IG users in, 30
Jarre, Jérôme, 112
Ja Rule, 239
JavaScript, 6
@jayzombie, 42
Jenner, Kendall, xix, 174, 186, 238–39
Jenner, Kris, xix–xx, 135, 137–38, 180
Jenner, Kylie, xix–xx, 161, 162–63, 174, 230, 245
- as youngest self-made billionaire, xx

Jobs, Steve, 65
Jolie, Angelina, 152
Jonas, Nick, 192
Jonas Brothers, 133
@JonBuscemi, 236
@jordandoww, 171
#jumpstagram, 104
Justice Department, U.S., 278

Kalanick, Travis, 23
Kaplan, Joel, 211
Kardashian, Khloé, xix
Kardashian, Kourtney, xix
Kardashian-Jenner family, 129, 230, 231, 264
- IG as main branding tool of, 135, 137–38

Kardashian West, Kim, xix, 47, 135–36, 137–38, 139, 180, 218, 230, 244–45

Kattan, Huda, 247
Keeping Up with the Kardashians, xix, 135, 137, 218
Kelly, Drew, 81–82
Kendrick, Anna, 148
@kevinbrennermd, 244
Keys, Alicia, 203
Khan, Imran, 200–201
Kicksta, 246
#kindcomments, 250
King, Nate, 185, 186, 205, 206
Klip, 109
Kloss, Karlie, 156, 218
Knowles-Carter, Beyoncé, 230
Koum, Jan, 125, 256
Kramer, Julie, 232
Krieger, Mike, xvii, xxii, 37, 50–51, 55, 63, 69, 76, 105, 140, 219
- Brazilian childhood of, 12
- Cox and, 257
- Crime Desk SF app of, 12, 32
- as disillusioned with FB's grow-at-all-costs culture, xvii
- in effort to preserve IG's brand, 176, 209
- at Golden Globe Awards, 192
- IG posts by, 31
- in increasing conflict with FB, 95–96, 214, 262–63, 271
- leadership philosophy of, 18
- at Meebo, 12
- philanthropy of, 72
- post-IG, 277
- problem solving by, 18, 30–31, 32, 33, 38–39, 110
- rectangular photo format and, 177
- resignation of, from IG, xxii, 272–75
- simplicity valued by, 18, 20, 21, 27, 102, 119, 191, 255
- Snapchat Stories and, 188, 192–93

Systrom's relationship with, 11–12, 13, 16–17, 33–34, 107, 254
Zuckerberg's meeting with, 60
Zuckerberg's relationship with, 252–53, 254–55, 256, 264
Kushner, Joshua, 45, 70, 218
Kutcher, Ashton, 44–46, 148, 172, 229
Systrom's friendship with, 46, 133

Lady Gaga, 158, 192, 204
@ladyvenom, 142
Lafley, A. G., 191
Lal, Arvin, 145
Lam, Lucian Yock, 169
Lawrence, Jennifer, 152
Lee, Aileen, 61
Lee, George, 177–78
Leibovitz, Annie, 186
Lester, Jason, 82
Leto, Jared, 44, 152
Levine, Marne, 180, 204–5
LeWeb tech conference, 83
Leyes, Mark, 99
Li, Alex, 187–88, 189, 192
Libra, 257, 263
likes, xvii, xxi, 20, 31, 39, 128, 144, 158, 169, 173, 175, 233, 236, 241, 242, 250, 251, 275, 279
Lincoln Center, 131–32
Linkin Park, 128
Lirag, Rafael, 85
Lord & Taylor, 235
Louis Vuitton, 119
Lovato, Demi, 192
#lowdownground, 104
Lowercase Capital, 23

McAllister, Philip, at IG, 51, 53, 73
McFarland, Billy, 238–39
McGill University, 129
McKinsey, 241, 242

Madonna, 126, 129, 133
Major Lazer, 238
Mandelbaum, Fern, 2–3
Marcus, David, 257
Marooney, Caryn, 202, 222
Marshalls, 4
Mase, 174
Mashable, 84
Matsubayashi, Kohji, 97–98
Mattel, 136
Mayer, Marissa, 8
media, *see* news media; social media
MediaKix, 236
Meebo, Krieger at, 12
Meekins, Marlo, 112
memes, 170, 181, 229, 251, 261
Meow Wolf, 242–43
mergers & acquisitions technology, 48, 64
message boards, online, 129
Messina, Chris, xx–xxi
Met Gala, 264
Metzger, Darwyn, 111–12
Michael Kors, 119, 120
Michele, Alessandro, 195
Michigan, University of, 247
Milan Fashion Week, 195
Ministry of Foreign Affairs, North Korean, 82
misinformation, *see* fake news
#MKTimeless, 119
mobile apps:
IG designed as, 37; *see also* Instagram
smartphones and, 10
Systrom's early efforts at building, 9–10
for video, 109–10
see also specific apps
mobile phones, xvii, 56–57
FB advertising on, 105, 150

mobile phones (*cont.*)
  photo technology of, xxi, 18,
    167–68, 191
  see also iPhones
monopoly, monopolies, 71, 75, 278
Monster.com, 3
Moonsplash, 4
Moore, Jason, 136–37
Mosseri, Adam, 211, 263–64, 271,
    272–74, 275, 278, 279
Mosseri, Monica, 273
Murphy, Bobby, 113, 115, 116
Muse, Arizona, 156
Museum of Ice Cream, 242
Musk, Elon, 22
Myspace, 1, 27, 183, 184

Nasdaq, 74
National Center for Missing and
    Exploited Children, 42
National Geographic, 153, 169
National Portrait Gallery, London,
    148, 166
National Public Radio:
  IG account of, 35
  "Status Update" episode on, 188–90
National Snake Day, 218
Nayak, Priya, 182, 183
Nest, 64
Netflix, 79, 80
Netscape, 11
Nevo, Aviv, 66
News Corp, 24
news media, 37, 69, 80, 82, 85, 86,
    142–43, 149, 155, 161, 171, 192,
    210, 240, 265, 274, 275
  celebrities' manipulation of, 136,
    138
  fake, *see* fake news
  FB and, 148–49, 191, 210–12, 250,
    257–58, 262, 264

IG and, 34, 35, 89, 99, 145–46,
    154–55, 174, 235, 251, 262, 264,
    271, 275
  and IG verified accounts, 231–32
  Snapchat and, 114, 192
  Systrom's relationship with, 33, 125,
    223
  Twitter and, 151–52
  see also social media
New York, 280
@newyorkcity, 43, 48, 82
New York Fashion Week, 131–32, 146
New York Times, xviii, 5, 43, 78, 86, 98,
    148, 208, 258, 273
New York University, 6, 43
Nextstop:
  sold to Facebook, 32
  Systrom at, 9–10, 11
Niche, 165
Nike, 82, 167
Ning, 127, 130
Ninja, 265
#nofilter, xxi, 155, 163, 248
North Korea, IG user in, 81–82
notifications, 90, 93, 102, 177, 270, 277
  push, 96, 97, 220
nudity, 97, 249
Nyong'o, Lupita, 152
Nyong'o, Peter, 152

Obama, Barack, 213, 224
  IG account of, 47
  Twitter account of, 47, 126
Obama, Michelle, 47
Observer, 258
Oculus VR, 64, 217, 253
Odeo, 6, 7, 8, 17, 27
  Systrom at, 5–6, 7, 12
  Twttr launch of, 7–8; *see also*
    Twitter
Office of Fair Trading, U.K., 76, 78

Oklahoma University, 144
Olivan, Javier, 92, 260, 268
Oliver, Jamie, 81, 147–48
Olsen twins, 136
Olson, Parmy, 256
Olympics, Summer (2016), 203
Omnicon, 121
Onavo, 122, 123, 124, 181
opioids, 261–62
Oprah's Book Club, 138
Orrick, Herrington & Sutcliffe, 75
Oseary, Guy, 44–45, 126, 133, 134, 138, 148
Ostrovsky, Josh, xx
"outfit of the day," 81
Owen, Andrew, 153, 203–4
Oxford English Dictionary, 152
OxyContin, 261

Palo Alto, Calif., 57, 60, 64, 68, 94
    FB office in, 15, 88, 107
    tech companies in, 5
Pandora, 10
Paris Fashion Week, 186
Parker, Sean, 2
passwords, 32, 48, 175
Path, 33, 76, 77
Payr, Marion, 142–43
Pelé, xv
Penguin Random House, 142
Penner, Carolyn, 84
Pepsi, IG account of, 35
PerezHilton, 136
Perino, Dana, 208
Periscope, 64
Perle, Liz, 154, 161, 170–72, 182
Perry, Katy, 127
"personal paparazzi," xvi
Pfeiffer syndrome, 161
Phantom, 111
Photobox, 2, 24

photographs:
    editing tools for, 243, 245
    filter apps for, *see* filter apps, photo
    rectangular format for, 176–77
    square format for, 5, 19, 22, 31, 33, 110, 147, 176
photography, Systrom's passion for, 2, 4–5
photo-sharing apps, 33
    see also Instagram
Picaboo, 113
Pichai, Sundar, 225
PicPlz, 33, 36, 76–77
Pinterest, 45, 79, 80, 172
Pitt, Brad, 152
Pixable, 76
Pixar, 180, 182
pixels, 19
Plans, 18
Playboy, IG account of, 35
Playing to Win (Lafley), 191
podcasts, 5, 237
pods, 246
Poke app, 115–16, 124, 191
Polaroid, Polaroids, 22, 31, 35
Poler, 229–30
Pons, Lele, 110, 112, 184, 265
Porch, Charles, 129–30, 139, 140, 143, 146, 149, 150, 154, 155, 157–58, 160, 171, 195–96, 218–19
    celebrities wooed to IG by, 126, 128–29, 130–35, 148, 158, 195–96
    fashion community, 131–32, 146
    as head of partnerships at IG, 160, 171, 219, 230
    in move from FB to IG, 128–29, 130–31
    Randi Zuckerberg and, 126–28, 130
    Systrom and, 130–31, 132–33, 135, 147–48, 166, 195, 245
pornography, on internet, 41

Portugal, 12
Prada, Miuccia, 195
press, *see* news media
Princeton, N.J., 129
Procter & Gamble, 191
#promposal trend, 170
Purdue Pharma, 261
Pusha T, 238
push notifications, 96, 97

quantitative analysis, 8
Quarles, James, 165
Quora, 24

Ranadive, Ameet, 271
"Rapture, The," 174
Ratajkowski, Emily, 238
Ray, Hannah, 166–67, 168
Reddit, 246
Red Hot Chili Peppers, 129
re-gram button, 45, 102
Regrann, 140
relevance, xxii, 65, 130, 195, 204, 276,
        280
    celebrities' concern with, 130, 204
    gaming the system and, 231–32
    teens' concern with, 171, 188–90
    Zuckerberg's concern with, 65, 124
Renaissance, 3, 4
Repost, 140
re-sharing, 80, 140, 157
    disallowed on IG, 43, 44, 140, 157
    viral, 20, 43–44, 140, 152, 210
Reuters, 54
Rich, Jessica, 235
Richardson, Bailey, 48, 81, 153
Riedel, Joshua, 32, 37, 43, 52, 53, 102,
        153
    Instameets organized by, 34–35,
        39–40
Rihanna, 47, 128, 218, 232

Rise, Cole, 34–35, 39, 40
    filter apps developed by, 23, 35,
        245
    as IG beta-tester, 22–23
Roberts, Julia, 152
Rolling Stone, 158
Rolling Stones, 148
Ronaldo, Cristiano, 230
Rose, Dan, 50, 95, 100, 200–201
Rosen, Guy, 260, 261–62
Rousteing, Olivier, 186
Rowghani, Ali, 49, 55
Royal Society for Public Health
        (RSPH), U.K., 248
Rubin, Dan, 30
Rusli, Evelyn, 118
Russell, Molly, suicide of, 277–78
Russia, fake news propagated by,
        224–25, 235, 250, 255
Russian Internet Research Agency
        (IRA), 251

Sacca, Chris, 23–24, 32
Samsung, 82, 152
Samsung Galaxy Note, 82
Sandberg, Sheryl, 123, 194
    as FB COO, 58, 100–101, 118, 121,
        211, 225, 258, 272
    and Zuckerberg's attempt to buy
        Snapchat, 200–201
San Francisco, Calif., 11
São Paulo, Brazil, xv, 13
Saverin, Eduardo, 15
scale, xviii, 37, 98, 103, 143, 201, 225,
        241, 265
Schake, Kristina, 222
Schrage, Elliot, 211
Schroepfer, Michael, 62, 95, 214, 225
Schuetz, Nicole, 20–21, 185–86, 254
Schultz, Alex, 223, 227
Scoble, Robert, 54

Scotch app, 19

Seacrest, Ryan, 135

Seagram, 12

Securities and Exchange Commission (SEC), 150

selfies, 114, 152, 161, 170, 182, 186, 188–89, 192, 208, 237–38, 240, 245, 247

"Selfies—Living in the Era of Filtered Photographs" (article), 244

Seliger, Mark, 158

Senate, U.S.:
    Intelligence Committee of, 225, 251
    Zuckerberg's testimony to, 258
    see also Congress, U.S.

Sequoia Capital, 55, 56

sexting, 114

sharing, xvi, 20, 45, 90, 92, 140, 147, 162
    ephemeral, 191, 193, 195, 214, 215–16, 217
    as key to FB's growth, 92, 94, 162
    of personal data, 16, 77, 135
    viral, 140, 152
    see also re-sharing

Sharp, Nathan, 199, 202

Sheng, Leo, 161

Shredz, 145

Siegler, M. G., 34, 36

Silicon Valley, 3, 4, 6, 8, 9, 15, 38, 44, 52, 56, 74, 172, 231, 284
    apps race in, 10
    buzzwords in, 18
    and FB's acquisition of IG, 59, 61
    mobile revolution in, 27
    salaries in, 71
    startup success rate in, 29
    talent wars in, 88

Simo, Fidji, 176, 254–55

Simple Life, The (reality TV show), 136

simplicity, Systrom and Krieger's philosophy based on, 18, 20, 21–22, 24, 119, 255

Skinny Confidential, 237

Smalls, Joan, 156

Snapchat, 122, 123, 124, 157, 171, 187, 198, 199–200, 217, 218, 223, 248, 277
    celebrities' use of, 192, 204
    cool factor of, 116
    ephemeral content on, 112–13, 114, 117, 216
    founder of, *see* Spiegel, Evan
    growth rate of, 216
    IPO of, 227
    reverse chronological order of posts on, 117
    teens' use of, 183
    as threat to IG, 123, 178, 181, 184, 192–93, 201, 223, 227
    Zuckerberg's attempts to buy, 114–15, 116, 117, 122, 125, 183, 191, 200–201

Snapchat Stories, 117, 122, 123, 191, 214
    ephemeral content of, 185
    IG's copying of, *see* Instagram Stories
    Systrom and, 188, 190, 192, 202–3, 217
    teens' use of, 183

Snap Inc., 227

Snoop Dogg, 4, 71, 98, 138
    early IG account of, 35–36

Socialcam, 109

social media
    as amplifier of issues, 278
    as both reflection and modifier of user behavior, 233, 234–35
    bullying on, 40, 41, 135, 161, 163, 218–19, 248

social media (*cont.*)
  FB's dominance of, 88, 121, 124,
    151, 209
  user passivity on, 234
  see also news media
social media companies, xvii, 109–10,
  203, 232
  see also specific companies
Social Network, The (film), 15, 67, 107
social networks, 88
  follower-based vs. friend-based,
    20, 80
  interest-based, 20, 21, 210
  virality and, *see* virality
Sony, 167
South by Southwest technology
    conference, 55
#sp, 236
Spacey, Kevin, 152
Spain, IG in, 226
spam, 80, 226, 260
Spectra photo filter, 23
Spiegel, Evan, 112–14, 115–16, 123,
    179, 191, 194, 195, 199–200
  Zuckerberg and, 116–17, 200,
    201–2
  see also Snapchat; Snapchat Stories
Spotify, 45
Square, 15, 46, 65
Squires, Jim, 120
Stanford Mayfield Fellows Program,
    5, 12, 46
Stanford University, 1, 2, 10, 11, 12, 20,
    24, 47, 173
Starbucks, IG account of, 35
startups, xx, 14
  see also specific startups
State of the Union address (2012), 47
status updates, 1, 12, 15
Stein, Robby, 194, 201
storytelling, xviii

@strawburry17, 171
Streep, Meryl, 152
Stretch, Colin, 225
Stuart Weitzman, xix
Styles, Harry, 130, 133
suicide, suicidal content, 40, 41, 42,
    270, 277–78
Sun, Fei Fei, 156
Sutro photo filter, 23
Swain, David, 154
Swank, Hilary, 192
Sweeney, Shayne, 32, 37, 53, 71, 75–76,
    79
Swift, Taylor, 47, 131, 204, 217, 218–19,
    231
Syracuse University, 232
Systrom, Diane, 3
Systrom, Doug, 3–4
Systrom, Kate, 3, 192
Systrom, Kevin, xxii, 31, 69–70, 94–95,
    123–25, 146, 153, 159–60, 180,
    193, 225, 245, 260, 261, 277
  at Academy Awards, 191–92, 204
  analytics and, 226–27
  art history and Renaissance as
    interests of, 3, 106
  celebrities' relationships with, 46,
    133–34
  childhood of, 3–4
  Clinton and, 207–8
  competitiveness of, 107–8
  Cox and, 257, 267–68, 272
  cycling by, 185, 186, 205–6
  deejaying by, 4, 10
  disillusioned with FB's grow-at-all-
    costs culture, xvii
  Dorsey and, 6–7, 15–16, 19, 60, 84
  early mobile websites built by,
    9–10; *see also* Burbn
  early prediction of IG success by,
    29

in effort to preserve IG's brand,
159–60, 176, 177–78, 184–85,
209, 217–20
in Florence, 3, 4–5, 19, 21
and FTC investigation of IG sale to
FB, 75, 76
at Google, 8–9, 23, 37, 58, 62, 194
IG founded by, *see* Instagram
IG posts by, 31
IG sold to Facebook by Krieger
and, *see* Facebook, Instagram
acquired by
IGTV and, 257, 264, 265–66, 267
in increasing conflict with FB, 214,
249, 252–53, 262–63, 268–69
Krieger's relationship with, 11–12,
13, 16–17, 22–23, 33, 107, 254
Kutcher's friendship with, 46, 133
leadership philosophy of, 18
at Middlesex boarding school, 134
Monday leadership meetings of,
107
at Nextstop, 9–10, 11
at Odeo, 5–6, 7, 12
1 million followers of, 187
perjury allegations against, 86,
98–99
photography passion of, 2, 4–5
Pope Francis's meeting with,
195–96
Porch and, 130–31, 132–33, 135,
147–48, 166, 195, 245
post-IG, 277
as pressured by Zuckerberg to build
IG's business model, 163–65, 167
problem solving by, 18, 32
as public face of IG, 33
re-sharing disallowed by, 140
in resignation from IG, xxii, 272–75
similar background of Zuckerberg
and, 106–7

simplicity valued by, 18, 20, 21, 24,
27, 45, 54, 102, 119, 160, 180,
189, 191, 199, 205
Snapchat and, 188, 190, 192, 202–3,
217
at Stanford University, 7, 8, 24
Stories opposed by, 190, 191
study abroad of, 3–5
on *Tim Ferriss Show*, 87
#trashcangate and, 181–82
well-being initiative of, 249
Zuckerberg's 2005 meeting with,
1–3
Zuckerberg's relationship with, 7,
38, 95, 104–5, 107–8, 163–64,
216–17, 251, 252–53, 256, 262,
264, 266–68, 269–70

tagging friends, 7, 90
tagging photos, 95
TaskRabbit, 17
tastemaking, tastemakers, xviii, xxi,
144, 237
see also influencers
Tatum, Channing, 149–50
Tatum, Everly, 149
#taylorswiftisasnake, 218
#tbt, 155
TechCrunch, 34, 35, 36
technology industry, 28
teens, 243
on FB, 117, 184
finsta accounts of, 182–83, 184
on IG, 118, 170
IG's analytic tools used by,
275–76
IG Stories and, 203
as key to the future of IG, 154, 171,
184
and pressure to post the best, 114,
170, 172, 188–90, 248

teens (*cont.*)

  Snapchat and, 115

  technology use by, 114

  unspoken social rules among, 182, 184

  Zuckerberg's resolve to better understand, 116

Teigen, Chrissy, 243

Telegram, 246

terrorism, terrorists, 249, 261

Tesla, 22

That '70s Show, 44

TheFacebook.com, 1, 7

  growth of, 2

  see also Facebook

@thefatjewish, xx

@theskinnyconfidential, 237

Thiel, Peter, 191, 193

#thinspiration, 41

This American Life (NPR show), 188

Threadsy, 17

Thrive Capital, 66, 70

Throwback Thursday, 155

TikTok, 277

Timberlake, Justin, 203, 204

Time, 38–39

Tim Ferris Show, 87

Tinder, 19

TMZ, 136

Toffey, Dan, 52, 53, 73, 141, 143

Totti Candy Factory, 242

Toy Story (film), 180

Transocean Ltd., 113

#trashcangate, 181–82, 204

travel, IG's influence on, 169, 241, 242

Trigger, Kaitlyn, 79

trolls, internet, 41, 219, 251

Trump, Donald, 207, 208, 210, 211, 224, 258

  FB leveraged by, 212–13

Trump, Ivanka, 70

Trump International Hotel, 70

Tumblr, 19, 103, 170

Tuna (dog), 141–42, 153

Tuna Melts My Heart (Dasher), 142

24 Hour Photo, 117

Twitter, xviii, xxi, 9, 17, 19, 31, 39, 130, 137, 151, 160, 170, 192, 225, 232, 239, 248

  Academy Awards and, 151, 204

  in attempt to buy IG, 25, 46, 48–49, 55–56, 86, 109

  Benchmark Capital investment in, 36

  chronological order of posts on, 19, 117

  content policing and, 43

  Dorsey at, 14, 25–26, 46–47

  early investors in, 23

  fake news on, 225

  as follower-based network, 20

  founders' discord at, 14

  free speech ethos at, 37, 156–57

  growth rate of, 216

  IG blocked from access to, 84, 99

  IG photo sharing to, 37

  IPO of, 98, 148, 149, 150–51

  Niche acquired by, 165

  Obama's account on, 126

  140–character limit of, 110, 128

  Periscope acquired by, 64

  retweet button of, 20, 44, 152, 157, 234

  status updates at, 15

  #tweetups and, 34

  as unwilling to edit content, 220

  user anonymity on, 41

  verification on, 132

  Vine acquired by, 64, 109, 157

Williams at, 14, 46
Zuckerberg's attempted purchase
    of, 57
Twttr, 7
Tycho (Scott Hansen), 34
Tyga, 238

U2, 126
Uber, 36, 45, 222
UberCab, 23
Underwood, Teddy, 120–21
"unicorns," 61

Van Damme, Tim, 51–54, 73
Vanity Fair, 158, 192
#vanlife, 229
venture capitalists, 2–3, 11, 15, 24,
    36–37, 55, 56, 109, 116, 191
Vergara, Sofia, 236
Verge, 216
verification, 231–32, 279
    as status symbol, 132–33
Verrilli, Jessica, 46
VidCon, 219
Viddy, 109
Vine, 64, 109–10, 111–12, 122, 124,
    157, 165, 171, 265
violence, violent content, 40, 41–42, 97,
    223, 249, 261
virality:
    of fake news, 225
    on FB, 162, 209, 211, 215, 251,
        260
    re-sharing and, 20, 43–44, 140, 152,
        210
    risky behavior and, 240, 243–45
    sharing and, 140, 152
    social networking and, 44
    on Twitter, 151, 239
Vogue, 118, 195, 231
    IG-related cover of, 156, 157

VPN (virtual private network), 122

Wall Street, 74, 102, 150, 151, 164, 266,
    267
Wall Street Journal, 102, 118, 122
Warner Bros. Records, 129, 130
Warr, Andy, 199
Washington Post, 208, 271
watchLAB, 182, 199
Watts, Kristen Joy, 153–54
Web 2.0, 5, 10
Weil, Kevin, 194–95, 263
West, Kanye, 218
WhatsApp, 64, 214, 256
    business plan lacked by, 256
    encryption on, 125, 256, 262
    FB's acquisition of, 125, 202, 217,
        255
WhatsApp Status, 216, 256
White, Emily, 100–101, 104, 105, 118,
    120, 122–23, 165
    in move to Snapchat, 123, 200
White House Correspondents' dinner,
    171
Whole Foods, 64
Wikipedia, 11
will.i.am (William Adams), 128
Williams, Evan, 5–6, 82
    Dorsey and, 25–26
    at Odeo, 27
    at Twitter, 14, 46, 57, 157
Winfrey, Oprah, 131, 136, 158
Wintour, Anna, 131, 156, 195, 264
WordPress, 17
World Economic Forum, 127
World Travel and Tourism Council, 241

X-Pro II filter, 21

Yahoo!, 2, 8, 11, 24, 116
    in attempt to buy Facebook, 57

Yahoo! (*cont.*)
  Flickr acquired by, 54
@yock7, 169
Young, Neil, 129
YouTube, 39, 59, 109, 112, 130, 137,
    157, 165, 171, 215, 219, 232, 240,
    248, 253
  algorithm on, 233–34
  Communications Decency Act and,
    41
  fake news on, 225
  Google's acquisition of, 53, 105
Yuki, Ashley, 175–76, 177

Zappos, 105
Zazzle, 5
Zero to One (Thiel), 191, 193
Zimmerman, Joel Thomas
    (Deadmau5), 70
Zipcar, 3
Zollman, Jessica, 40, 41–43, 53, 70,
    72–73, 97, 104
Zoufonoun, Amin, 58–59, 60, 61–62,
    84–85
Zuckerberg, Mark, 27, 52, 106, 112,
    148–49, 150, 213, 220–21, 262,
    274, 276
  in attempt to buy Twitter, 57
  in attempts to buy Snapchat,
    114–15, 116, 117, 122, 125, 183,
    191, 200–202
  better understanding of FB users as
    goal of, 221–22
  and Cambridge Analytica scandal,
    258
  as committed to IG's independence
    at FB, 54, 55, 63, 67, 95, 96, 106,
    121
  competitiveness of, 108, 148
  credit for IG's success taken by,
    266–67

in decision to buy IG, 56–57;
    see also Facebook, Instagram
    acquired by
  dominance as goal of, 78, 109,
    124
  and fake news scandal at FB,
    211–12, 225
  "family of apps" bundled together
    by, 255–56, 267–68
  FB support for IG shut down by,
    268–69
  and FTC investigation of FB's IG
    acquisition, 76
  Greek and Roman history interest
    of, 106
  hoodie worn by, 2, 74
  IG cannibalization of FB concern
    of, 223, 226, 227–28
  at IG video launch, 111
  Krieger's meeting with, 60
  Krieger's relationship with, 252–53,
    254–55, 256, 264
  and Krieger's resignation from IG,
    272
  in negotiations with Systrom, 57–58,
    60–62, 73; *see also* Facebook,
    Instagram acquired by
  new year's resolutions of, 221–22
  personal security of, 107, 134, 222
  problems anticipated by, 149,
    214–15
  rising tensions between Systrom
    and, 217, 252, 262, 267
  similar background of Systrom and,
    106–7
  6,000 word manifesto of, 221
  Spiegel and, 116–17, 200, 201–2
  Systrom pressed to build IG's
    business model by, 163–65,
    167
  Systrom's 2005 meeting with, 1–3

Systrom's relationship with, 7, 38,
55, 95, 104–5, 107–8, 216–17,
251, 252–53, 256, 264, 266–68,
269–70
and Systrom's resignation from IG,
272
testimony to Congress by, 258–59,
262

understanding of teens as goal of,
116
and WhatsApp, 64, 125, 217, 256
see also Facebook
Zuckerberg, Randi, 126–28, 130,
148
ZuckPri, 95
Zwift, 186